Computational Design
Technology, Cognition
and Environments

Rongrong Yu
Lecturer, Griffith Centre for Design and Innovation Research
School of Engineering and Built Environment
Griffith University, Australia

Ning Gu
Professor in Architecture
University of South Australia, Australia

Michael J. Ostwald
Associate Dean of Research and
Professor of Architectural Analytics
University of New South Wales, Australia

CRC Press
Taylor & Francis Group
Boca Raton   London   New York

CRC Press is an imprint of the
Taylor & Francis Group, an **informa** business
A SCIENCE PUBLISHERS BOOK

First edition published 2021
by CRC Press
6000 Broken Sound Parkway NW, Suite 300, Boca Raton, FL 33487-2742

and by CRC Press
2 Park Square, Milton Park, Abingdon, Oxon, OX14 4RN

Library of Congress Cataloging-in-Publication Data

Names: Yu, Rongrong, 1981- author. | Gu, Ning, 1975- author. | Ostwald, Michael J., author.
Title: Computational design : technology, cognition and environments / Rongrong Yu, lecturer, Griffith Centre for Design and Innovation Research, School of Engineering and Built Environment, Griffith University, Qld, Australia, Ning Gu, professor in architecture, University of South Australia, Australia, Michael Ostwald, associate dean of research and professor of architectural analytics, UNSW, Sydney, Australia.
Description: First edition. | Boca Raton : CRC Press, Taylor & Francis Group, 2021. | "A science publishers book." | Includes bibliographical references and index. | Summary: "With the rapid emergence and adoption of new computational design technologies in the design field, it is important to critically understand how designers response to those new environments. This book systematically explores the impact of emerging computational design environments on design and designers. It offers an unique opportunity to look into design thinking in the current digital age"-- Provided by publisher.
Identifiers: LCCN 2021004499 | ISBN 9780367203061 (hbk)
Subjects: LCSH: Computer-aided design. | Engineering design.
Classification: LCC TA345 .Y83 2021 | DDC 670.285--dc23
LC record available at https://lccn.loc.gov/2021004499

ISBN: 978-0-367-20306-1 (hbk)
ISBN: 978-0-367-77493-6 (pbk)
ISBN: 978-0-429-26078-0 (ebk)

Typeset in Palatino Roman
by Innovative Processors

Acknowledgements

The authors are grateful for the support and contributions from our extended team of professional colleagues, associates and assistants, who have all made the road to this research destination possible. The following co-authors have helped us to develop several ideas and projects contained in this book and we fully acknowledge their contributions: Professor John Gero, Professor Mary Lou Maher, Dr Peiman Amini Behbahani, John Wells and Professor Mi Jeong Kim. In addition, this book was proofread by Maria Roberts.

Some sections of this book are based on materials, projects and data that were previously published in journals, books and conference papers. Chapter 2 includes material adapted from: Gu N., Yu R., and Behbahani P. A., (2018), Parametric Design: Theoretical Development and Algorithmic Foundation for Design Generation in Architecture, by B. Sriraman (Ed.), *Handbook of the Mathematics of the Arts and Sciences*, Springer. Chapter 3 includes expanded and revised work based on: Yu R., Gu N., Ostwald M., and Gero J. (2015), Empirical Support for Problem-Solution Co-Evolution in a Parametric Design Environment, *Artificial Intelligence for Engineering Design, Analysis, and Manufacturing (AIEDAM)*, Vol. 29, Issue 01; Gero J., Yu R., and Wells J. (2019), The Effect of Design Education on Creative Design Cognition of High School Students, *International Journal of Design Creativity and Innovation*, Vol. 7 Issue 04; Yu R. (2014), *Exploring the Impact of Rule Algorithms on Designers' Cognitive Behaviour in a Parametric Design Environment*, PhD Thesis, University of Newcastle, Australia. Yu R., Gu N., Ostwald M., and Gero J. (2015), Empirical Support for Problem-Solution Co-Evolution in a Parametric Design Environment, *Artificial Intelligence for Engineering Design, Analysis, and Manufacturing (AIEDAM)*, Vol. 29, Issue 01 and Yu R. and Gero J. (2018), Using eye-tracking to study designers' cognitive behaviour while designing with CAAD, *52nd International Conference of the Architectural Science Association (ASA 2018)*, Melbourne. Chapter 4 includes materials adapted from: Yu R. (2014), *Exploring the Impact of Rule Algorithms on Designers' Cognitive Behaviour in a Parametric Design Environment*, PhD Thesis, University of Newcastle, Australia; Gu

N., Kim M. J., and Maher M. L., (2011), Technological Advancements in Synchronous Collaboration: The Effect of 3D Virtual Worlds and Tangible User Interfaces on Architectural Design, *Automation in Construction*, Vol. 20, Issue 03; Gu N. (2015), Generative Design Grammars: An Intelligent Approach Towards Dynamic and Autonomous Design, in: K. William's and M. J. Ostwald's (Eds.) *Architecture and Mathematics from Antiquity to the Future, Volume II: The 1500s to the Future*, Springer; Gu N., and Maher M. L. (2014), *Designing Adaptive Virtual Worlds*, De Gruyter; Yu R. and Gero J. (2018), Using eye-tracking to study designers' cognitive behaviour while designing with CAAD, *52nd International Conference of the Architectural Science Association (ASA 2018)*, Melbourne; Yu R., Ostwald M., and Gu N. (2015), Parametrically Generating New Instances of Traditional Chinese Private Gardens that Replicate Selected Socio-spatial and Aesthetic Properties, *Nexus Network Journal: Architecture and Mathematics*, Vol. 17, Issue 03; Yu R., Gu N. and Ostwald M. (2016) The mathematics of spatial transparency and mystery: using syntactical data to visualise and analyse the properties of the Yuyuan Garden. *Visualisation in Engineering*. Vol. 4, Issue 4; and Yu R., Gu N., and Ostwald M. (2018), Evaluating Creativity in Parametric Design Environments and Geometric Modelling Environments, *Architectural Science Review*, Vol. 61 Issue 06.

We thank all the journal editors and reviewers for their contributions to the development of this book.

Preface

There is often a level of confusion when people first discover the field of 'computational design'. Does it describe the process of designing computers, or alternatively the process wherein computers design artefacts? It is certainly not the former, which is more akin to computer science, and while it has some similarities to the latter, it is a much richer and more diverse field.

There are two common ways of understanding the field of 'computational design'. First, it can be defined as a systematic way of thinking about or conceptualising the design process. This could be understood as a rigorous reasoning method founded in the timeless fields of mathematics and logic. The second definition of computational design encapsulates the hardware, software and systems that support the act of designing. These have developed rapidly over time to respond to the needs of designers. The scope of this book encompasses both of these definitions of computational design, as well as the overlap or interdependency between them. This is a particularly interesting way of approaching the topic of computational design, because it emphasises the interactions between a seemingly stable body of knowledge on the one hand, and an increasingly dynamic technological environment on the other.

Three distinct aspects of computational design are introduced and explored in this book: design technology, design thinking and design environments. The first is associated with the hardware, software and systems used to support design, and which might in turn shape the way designers work and think. The second is concerned with the cognitive mental processes that occur during acts of designing, and which are influenced by computational design technologies. The third explores a variety of coexisting computational environments and contexts that are available to designers. This book does not, however, completely separate these three. Instead, it examines the connections between them, gradually building up a picture – literally a multi-stage figure and conceptual model – which explains how they shape one another. For example, advances in technology create new environments for designers, which can in turn, facilitate new ways of design thinking. These new ways of thinking then require new environments to support them. Thus, between those three different aspects, there is not only overlap but direct influence. This is one of the motivations for the present book, to reveal the many relationships that exist between design technology, design thinking and design environments, and which also can support innovation or creativity.

Throughout this book a recurring observation is that design, as a computationally supported process, is undergoing a radical evolution. The 'stable' and 'timeless' logic of the design process is actually evolving, and sometimes in unexpected ways. This realisation is the catalyst for several important questions. In a fast changing world, which computational tools are most appropriate for advanced architectural and design practices? Do these advanced tools change the ways designers thinking and act? Are newly established computational design environments actually supporting or hindering creativity? What methods can we use to critically analyse the influence of computational design environments on both design processes and designers? These are critical questions which scholars and practitioners are asking. They can also be complex to answer, as technological developments and advances in research, will render some answers obsolete before they can be answered. As such, this book does not attempt to provide perfect answers, instead it offers the knowledge and conceptual tools to continue to ask these critical questions and aim to develop timely answers.

Rongrong Yu
Ning Gu
Michael J. Ostwald

Contents

Acknowledgements   iii

Preface   v

1. Introduction   **1**

  1.1. Computational design   1
  1.2. Design technology, cognition and design environment   3
  1.3. Summary of chapters   6
  1.4. Context for the book   7

2. Emergent Technologies in Computational Design   **10**

  2.1. Introduction   12
     2.1.1. History of CAD   12
     2.1.2. Evolution of digital modelling devices   14
     2.1.3. Evolution of CAD software and hardware   16
  2.2. Parametric and generative design   20
     2.2.1. Key concepts and characteristics of generative design   21
     2.2.2. Major generative design systems   23
     2.2.3. Key concepts in parametric design   26
     2.2.4. History and application of parametric design   30
  2.3. Presence and telepresence, virtual and real worlds   35
     2.3.1. Telepresence   35
     2.3.2. Augmented Reality   36
     2.3.3. Virtual Reality   45
  2.4. Conclusion   48

3. Understanding Design Cognition in Computational and Generative Design   **58**

  3.1. Design cognition   59
     3.1.1. Design thinking   60
     3.1.2. Design problems and design solutions   61
     3.1.3. Design creativity   62

3.2. Formal approaches to studying design cognition   66
    3.2.1. Protocol analysis   66
    3.2.2. Biometric approaches to studying design cognition 72
3.3. Design cognition in the computational design environment 77
    3.3.1. Computational thinking and design thinking   77
    3.3.2. Design cognition in the computational environment 78
    3.3.3. Design cognition in the Parametric Design
        Environment (PDE)   80
3.4. Conclusion   83

4. Cognitive Impacts and Computational Design Environments   101

4.1. Introduction   101
4.2. Case study 1: Designers' behaviour in parametric and
    geometric design environments   104
    4.2.1. Research design   107
    4.2.2. Result 1: Design issues and processes   108
    4.2.3. Result 2: Designers' cognitive effort   111
    4.2.4. Result 3: Design patterns   115
    4.2.5. Result 4: Co-evolution process in parametric design 117
4.3. Case study 2: Cognitive studies of design collaboration in a
    virtual environment   125
    4.3.1. Collaborative design studies and technologies   126
    4.3.2. Experiments and coding scheme   128
    4.3.3. Protocol analysis results and discussion   129
4.4. Case study 3: A biometric approach to analysing cognitive
    behaviour in a CAD environment   131
    4.4.1. Experiment   132
    4.4.2. Results   132
4.5. Case study 4: Implementing rules in design, using generative
    design grammars   138
    4.5.1. Design grammars   138
    4.5.2. The conceptual framework of generative design
        grammars   139
    4.5.3. Design rules   140
    4.5.4. Designing a virtual gallery   144
4.6. Case study 5.1: Generating new design instances of an
    existing style using computational analysis   146
    4.6.1. Stage 1: Syntactical derivation   147
    4.6.2. Stage 2: Parametric generation   153
    4.6.3. Stage 3: Fractal analysis   157
4.7. Case study 5.2: Transparency and mystery in traditional
    Chinese private gardens   163
    4.7.1. Pedestrian accessibility convex map analysis   164
    4.7.2. Visual accessibility based isovist analysis   165

4.7.3. Hypothesis framing 166
4.7.4. Exploration of Yuyuan Garden's transparency
and mystery 166
4.8. Case study 6: Creativity in a parametric design environment 174
4.8.1. Research design 175
4.8.2. Analysis of results 176
4.9. Conclusion 184

5. Conclusion 197

5.1. A conceptual model 197
5.2. Looking into the future of computational design 204
5.2.1. Design technology: Implications and future
developments 204
5.2.2. Design cognition: Implications and future
developments 205
5.2.3. Design environment: Implications and future
developments 206
5.3. Conclusion 207

Appendix: Coding Example for Case Study 1 209

Appendix of Images Sources 238

Index 240

Introduction

1.1. Computational design

This book is about 'computational design', which is defined as both a computational and systematic way of thinking about the design process and a set of tools or techniques to support it. Specifically, this book traces developments in three aspects of computational design. The first is concerned with technology and its capacity to support and even revolutionise the design process. The second is about the cognitive or thought processes that occur while designing, especially when adopting new computational technologies. The latter examines the different design environments or contexts that now co-exist and can be used by designers. Rather than being separate themes, these are closely connected, as advances in technology create new environments for designers as these environments support new ways of thinking. These new ways of thinking also require new technology to support them. Thus, there is an overlap between these three themes, but more importantly, there is a direct agency or influence. One of the messages of the present book – is that design technology, thinking and environments are not isolated topics, and advances in each spur innovation in the others. This introductory chapter defines key terms, concepts and themes in the present book. Thereafter, it outlines the structure of the book, its intended readership and scope.

Design is a process that spans from ideation to realisation, or from the first concept to the final product. It can also be understood as a creative problem-solving process with a pragmatic purpose or goal. Between the creative idea and the pragmatic resolution, the design process is typically understood as operating in a series of stages. Each stage cyclically refines and tests ideas, gradually transforming the designer's initial vision into a functional outcome that can be manufactured or constructed. These intermediate stages in the process have various names that reflect increasing levels of definition, certainty and accountability. For example, in architecture and industrial design these stages are often known as 'concept design', 'schematic design', 'developed design'and 'detailed design'. Such names are necessarily artificial, as too are the number and sequence of the stages. In reality, every design process is slightly different, with some needing to repeat stages until an acceptable outcome is produced, and others truncating, merging or even skipping them. Nonetheless, this staged model of design offers a valuable framework for understanding a complex process. It is used in education, industry and across the design professions, and it is even embedded in contracts. It is, however, not the only way of understanding, undertaking or analysing design.

The design process can also be conceptualised in computational terms as an iterative operation with defined 'inputs', 'rules' and 'outputs'. The inputs might include the client's and user's needs, budgetary constraints or siting information. The rules typically require a formulation of performance expectations for the design, and the outputs are the completed documentation, specifications and approvals for the final product. Once again, the actual process taken by an individual designer may not follow this precise structure of inputs, rules and outputs. Novice designers, and those with singular visions, may have more idiosyncratic approaches that do not follow this model. For most designers and most purposes however, a computational model of the design process can provide deep, measurable insights and a formal structure as the catalyst for new technical and cognitive developments.

As the opening paragraph in this chapter reveals, computational design is typically defined as *both* a model of the design process and a technologically supported process for the same. As such, it incorporates two dimensions: an overarching sense of the cognitive logic of design, and a consideration of the tools used to support it. It must be acknowledged that some authors differentiate 'design computing' from 'computational design' to delineate these two properties more clearly. For example, they use 'design computing' to describe a process model that is framed in terms of logic structures and operations, and 'computational design' to refer to tools or systems that support the generation or automation of the process. This distinction is not universally accepted and it is not adopted in the present book. The phrases 'design computing' and 'computational design' have been used interchangeably in the past, and the two different aspects of their definitions are merged in the present book. The decision not to differentiate between the two was taken because this book addresses both computational models and tools and the overlap between them.

Considering just the first part of the definition of computational design, being concerned with a process model, a further clarification is needed. The word 'computational' refers to the use of mathematical logic or algorithmic systems for formalising parts of the design process. Contrary to first impressions, this does not mean that computer hardware and software are required. The computation may be undertaken using pen and paper, or graphic rules supported by simple numerical processes.

Indeed, the first three decades of research in this field were largely undertaken without computers. The message for the reader here is that when this book talks about computational models of the design process, it is referring to formal and systematic approaches, such as algorithmic models, that may or may not involve a computer.

The second part of the definition of computational design, pertaining to the way tools and technology support design, is also significant for the present book. Throughout history, designers have always relied on tools, techniques and protocols to complete their work. These can collectively be thought of as 'design enablers', and they include everything from pencils, rulers and paper to advanced software, multi-core processors and 3D printing. A common point of contention in design research revolves around the extent to which these enablers shape or influence the design process. For example, imagine a plan of an object drawn by hand on vellum and a plan of the same object drawn in a computer and printed on paper by an inkjet printer. One point of view holds that the enablers (pencil and vellum, or CAD and printer) are completely irrelevant, as the outcome is identical in terms of the information it contains and transmits. Another point of view is that the two are completely different, as the process of holding and sharpening the pencil, of creating a sequence of marks on the vellum and using pumice to correct any errors, necessarily involves an intimate relationship between the designer's hand and the drawing. Moreover, it is possible that the time taken to produce the drawing allows the designer to think through the choice of materials or colours needed. Conversely, the hand drawing only contains as much information as is needed for the specific task, whereas the CAD drawing contains many layers of embedded information that might describe the scale, materiality, construction sequence and even cost of the design. Thus, the computer-assisted version assumes that a higher degree of resolution may have occurred before the final drawing is extracted from a CAD view and exported to a window for being saved as a PDF and then printed. This simple example highlights multiple issues about immediacy, phenomenology and cognition in the design process, and the impacts of technology.

1.2. Design technology, cognition and design environment

This book accepts as a general premise that the choice of design tools or enablers necessarily has some impact on the cognitive behaviour of designers, and hence on the design process and the final design product. The level of impact then becomes the source of debate, experimentation and discussion. For example, in a technological sense, there may be a minor impact when changing from an HB to a 2B pencil, or from a rapidograph to a felt-tip pen. However, the change from two dimensional sketches mounted on a wall, to three-dimensional data visualisation in virtual reality is more pronounced. In terms of design thinking, the change from drawing a shape on a piece of paper to scripting a shape in a computer requires a paradigm shift in cognitive processes. From an environmental perspective, designers have always compared two or three sketches of a design to decide which is better. Today, designers can very rapidly generate hundreds of design options and program software to rank and select the best ones. This approach is called generative design or algorithmic design; it is based on rules or algorithms and it has opened up new design frontiers. This too, suggests that a major change is occurring in the design process.

Across these three themes of computational design – technology, cognition and environment – it is clear that the design field itself is often undergoing radical evolution. However, there has been a lack of critical and detailed analysis of the implications of this evolution and these new developments are raising multiple questions of significance. For example, what are the latest emerging computational tools that are applicable to architectural and design practices? Are we seeing significant changes in designers' ways of thinking as a result of these emerging tools? Do new computational environments ultimately support or hinder creativity? How can we critically analyse and assess the impacts of new computational design environments on design and designers? To address such questions, this book introduces three broad categories of enablers: technical tools, cognitive factors and environmental developments. It considers their applicability to architecture and design and explores their impact on both design and designers. This categorisation proposes a comprehensive, systematic and innovative review of these developments in computational design.

It is also important to define the three themes of computational design – technology, cognition and environment – which are core to this book.

1. 'Design technology' refers to both the technology embedded in a final product and the knowledge of processes, information and applications involved in developing a product (Bozeman, 2000). Therefore, design technology may be defined as the knowledge, applications and processes involved in developing a design. It can refer to both computational and non-computational (i.e. traditional) design technologies, but increasingly design technology refers to the former more than the latter. This is especially so considering the wider acceptance of computation in design as well as our contemporary industries and societies.

2. 'Design thinking' refers to the capacity to understand a person's design process or to effectively employ this process. This goal is the catalyst for the field of 'design cognition', which focuses on the

study of mental processes, strategies and knowledge areas employed whilst designing (Visser, 2004). For example, Cross (2001) defines design cognition as a cognitive science that studies problem-solving behaviour (including both problem finding and problem solving). Many cognitive studies seek to address the question, 'how do designers think?' Design thinking has been described as comprising a set of six primary processes, with some secondary variables involved, including formulation, synthesis, analysis, evaluation, documentation and reformulation (Gero, 1990).

3. An 'environment' is defined as 'the conditions that people live, work, or spend time in and the way that they influence how they (people) feel, behave, or work'. Within the computing domain, an environment is 'the system in which a computer or computer program operates' ("Environment", n.d). The 'design environment' is the set of conditions that affect a designer's way of working, including relevant computational operating systems and processes. The medium or environment in which the design is undertaken – be it physical and sketch-based or digital and CAD-based – has a significant impact on designers' cognitive processes (Chen, 2001; Mitchell, 2003).

The relationships between design technology, design thinking and design environment are initially modelled in Figure 1.1. This model is framed by the broader design environment. The Design Environment (D_{Env}) encompasses both the technology that supports and enables the design process and the cognitive operations and behaviours that occur in this process. As such, D_{Env} is the joint product of both the tools or enablers and the thought processes and related actions. Arguably, D_{Env} is greater than the sum of both technology and cognition, as it includes additional factors (i.e. systems) that are intrinsic to design operations. These might include quality assurance mechanisms, contractual conditions, documentation and archival systems, all of which are part of the environment but are neither enablers nor related to cognitive processes. Within D_{Env}, Design Technology (D_{Tec}) is the set of tools which enable the modelling, visualisation, analysis and generation of different components of the design. Design Cognition (D_{Cog}) within D_{Env} is the set of mental processes, behaviours and operations that occur during the design process. D_{Cog} is not entirely contained within the boundaries of D_{Env}, as there can be external influences outside the defined environment, on each person's cognitive operations. In summary, the combination of D_{Tec} and D_{Cog} make up the core of D_{Env}, but in practice, $D_{Env} > (D_{Tec} + D_{Cog})$ and D_{Cog} often stretches beyond D_{Env}. These relationships, as illustrated in Figure 1.1, provide a foundation for this book and they will also guide the development of a conceptual model to be presented in Chapter 5. The model provides a formal structure to guide designers in computational design practice through better understanding of design technology, design cognition and design environment. It is also useful as a guide for researchers and scholars engaged in critically reviewing current work or planning future developments in the field.

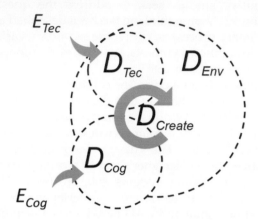

Figure 1.1. Relationships between design technology, design cognition and design environment.

1.3. Summary of chapters

Including this introduction, this book consists of five chapters. Chapters 2, 3 and 4 are each aligned to one of the three major themes of the book: design technology, design cognition and design environments. The final chapter draws together observations about all three themes and discusses their implications for computational design research and practice. This discussion is supported through the development of a conceptual model.

Chapter 2 introduces readers to emergent technologies in computational design. It discusses how these computational design technologies have evolved in response to new challenges or needs, and changed the way designers work. The purpose of this chapter is to provide readers with a broad understanding of these technologies, including their history, characteristics, applications and recent research about them. This chapter reveals that emerging computational design technologies have potentially changed the way designers work, and even the way they think.

Chapter 3 starts with defining and exploring design cognition in the context of computational design from three perspectives: design thinking, design problem and solution co-evolution, and design creativity. The first part of this chapter is built on a critical review of past research in the field, drawing on key literature from the 1970s to the present. The second part of Chapter 3 summarises and compares formal approaches to the study of design cognition, with a focus on cognitive and biometric approaches. This is followed by a discussion of design cognition in a specific type of computational design environment, the parametric design environment. Using a cognitive approach, the chapter shows that computational thinking in parametric design environments can have at least two levels – the design knowledge and algorithm rule levels. In parametric or generative design environments, designers not only use design knowledge, but also create explicit rules to achieve their design intentions. This chapter introduces and presents relevant theoretical underpinnings, formal approaches and methods, as well as specific examples of design cognition – a starting point for readers to better understand how designers think and behave in different computational design environments.

Building on the previous two chapters, Chapter 4 applies the theories and methodologies of design cognition in a series of case studies that examine the impact of different computational environments on design and designers. The first set of these case studies demonstrates the two formal methodological approaches introduced in Chapter 3. Protocol analysis, a cognitive approach, is employed to study designing in parametric design environments and in collaborative virtual environments. Eye-tracking technology, a biometric approach, is then used to understand the behaviour of designers in a CAD environment. The following set of case studies introduces and demonstrates two different computational design environments (based on shape grammar and space syntax) for design analysis and generation. The last case study focuses on support for creativity, as expressed in and compared between parametric design and geometric design environments.

Chapter 5 concludes the book by establishing a conceptual model for computational design, based on the key relationships between the three major themes of this book (as presented in Chapters 2 to 4 respectively). The development of the model includes five major steps, which are illustrated in Chapter 5 and supported by the findings of Chapters 2, 3 and 4. The model provides a formal framework for exploring the implications of computational design and forecasting new developments of future design technology, taking into account both design cognition and design environment. The goal of the model is to better support our understanding of design and designers. The conclusion to this book also discusses future trends for design technology development, and challenges and opportunities for designers in the computational era.

1.4. Context for the book

This book has been written primarily for a broad academic audience, including scholars, researchers, educators and postgraduate students interested in computational design. Some specialised undergraduate courses in design and computation might also benefit from the breadth and depth of materials covered herein. This book is also suitable for general audiences who have an interest in applying emerging computational technologies to design, such as professional designers (e.g. architects, urban designers, city planners and product designers). While some pre-knowledge of computational design and design cognition will be of assistance to better understand aspects of the book, core concepts are appropriately introduced in the text for general audiences. Each chapter is also fully referenced so that readers can find additional information or clarification to assist them further.

The focus of the book is on computational design concepts, technologies, practical applications and related research topics, and this book's authors possess a rich mixture of scholarly and professional experiences in architecture and design computing. As such, this book may discuss for example, programming and the intricacies of technological development, but it is not focused on these specific technical issues. Instead, its emphasis is on topics, themes and ideas that are of relevance to design and designers.

Ultimately, this book seeks to offer new insights about the impact of computational design on design and designers, through exploring the relationships between design technology, design cognition and design environment. New developments in computational design have been occurring so swiftly and decisively over the last decade, that many have not been critically and systematically studied and reported. For practical reasons, not every new computational design tool is included, and the case studies selected have an emphasis on computational design in architecture, even though the findings will often be valid for broader disciplines, including urban, interior and industrial design. The goal of this book is to advance contemporary design knowledge, and computational design in particular. It provides insights for understanding computational design and directing its future development.

References

Bozeman, B. (2000). Technology transfer and public policy: A review of research and theory. *Research Policy*, 29(4), 627–655. https://doi.org/https://doi.org/10.1016/S0048-7333(99)00093-1

Chen, S.-C. (2001). The role of design creativity in computer media. *Proceedings of the Nineteenth Education and Research in Computer Aided Architectural Design in Europe Conference*, Helsinki, Finland.

Cross, N. (2001). Design cognition: Results from protocol and other empirical studies of design activity. pp. 79–103. *In*: E. Charles, M. Michael and N. Wendy (Eds.), Design Knowing and Learning: Cognition in Design Education. Elsevier Science. https://doi.org/10.1016/b978-008043868-9/50005-x

Gero, J.S. (1990). Design prototypes: A knowledge representation schema for design. *AI Magazine*, 11(4), 26–36.

Mitchell, W.J. (Ed.). (2003). *Beyond Productivity: Information Technology, Innovation, and Creativity*. National Academies Press.

Visser, W. (2004). *Dynamic Aspects of Design Cognition: Elements for a Cognitive Model of Design*. (Research report) RR-5144, INRIA.

Emergent Technologies in Computational Design

In the modern industrial world, design and production are normally treated as separate processes, the assumption being that the design process has been completed before its production (manufacturing or construction) commences. Given the production costs and consequences of poorly conceived or completed designs, it is also assumed that a design has been thoroughly conceptualised, simulated and tested before assembly or manufacturing begins. However, throughout history the line between design and production was not always so defined.

In the ancient world, for example, the designer of small or standard objects or structures was often the producer. As such, there might have been no need for the designers to document their ideas, or preserve such documents if they existed at all, especially if the "documents" were reusable clay tablets or fragile models. Furthermore, the processes of designing and production may have been more interactive and responsive in the past. For example, architects and builders in ancient Greece and Rome constructed and tested full-size walls and structural sections as part of the design process, creating prototypes that may have been utilised in the completed buildings.

The earliest evidence that a separation existed between design and production is found in Mesopotamia, where people used tablets to inscribe the measurements of floor plans before starting construction (Donald, 1962; Tasheva, 2012). While it is not clear if they visualised design ideas or existing buildings consistently this way, they nevertheless created a level of basic documentation for objects or buildings on their papyri or inscriptions. Significantly, the orthogonal lines on those Mesopotamia tablets appear to necessitate the existence of an instrument for drawing them. Measuring instruments have been known and used since ancient times. The Indus Valley civilisation had rulers made of copper (Sundaram, 2017). Compasses were used across Asia and Africa, from China to Sardinia; Egyptians had wooden right angles and the Roman Empire used standard measures. However, these instruments may not have been solely developed for design; they may have also been used for trade, navigation, construction and standardisation. Regardless of their precise use or origins, by the medieval era such instruments were in common use by craftspeople and designers for conceptualising and documenting their work.

These early examples of design technologies could be classified into three main categories: instruments for modification (like carving awes or drawing styli), instruments for guiding (such as rulers, compasses and angles), and media like clay or paper to record the design during the process (Gu and Ostwald, 2012; Ostwald, 2012). Until the late 20th century, most of the tools used by designers fell into these three basic categories. Pens replaced graphite sticks or styluses, scale rulers and set-squares replaced rods and angles, and transparent tracing paper replaced animal skin and papyri. In general throughout history, such design technologies and tools mostly became more accurate, portable and affordable, rather than undergoing an essential shift in application. However, in the second half of the 20th century this situation changed completely. Computational design tools may have initially been conceptualised as digital versions of the three main categories of manual technologies, however they soon promoted a paradigm shift for designers. In the first instance, for example, they separated the displaying and recording functions of the medium, which were traditionally incorporated on the same page. They further made the automated modification of design possible. By associating parts or properties of design into objects, changing one object would automatically affect the rest in a controlled way. Finally, in recent years it has become possible to create immersive experiences to see or feel a design, in ways that were never possible or imaginable in the past. The technology and tools used by designers have changed, and the implications of this change for design thinking are still being understood.

This chapter describes the primary or mainstream computational design technologies currently used in the AEC (Architecture, Engineering and Construction) and design industries. Chapter 1 presents a background to the development of computer-aided drafting and Computer-aided Design (CAD) tools. These are the immediate successors of the manual technologies used throughout history. This chapter discusses how the various digital design technologies evolved in response to new pressures or needs, and changed the way designers worked. This is followed by a discussion of insights into design automation using programmatic and generative approaches to design. The final section in Chapter 1 reviews the new relationship between the design space and human environments enabled by the introduction of digital realities. The purpose of this chapter is to provide a basic understanding of computational design technologies, including their history, main characteristics, applications and recent research. This chapter builds the foundational knowledge for readers before they proceed to the remainder of the book.

2.1. Introduction

2.1.1. History of CAD

The origins of CAD are typically traced to the mid-1950s, when for the first time, early computers began to be used for more than just mathematical calculations or military observations. The expansion of computer applications into design became possible at this time as a result of three technological advances: improved memory usage, better human-computer interaction and enhanced mathematical formulation of geometry. The first of these developments was demonstrated in Lincoln laboratory's TX series of computers, which were purchased and developed by MIT in the 1950s (Weisberg, 2008). These systems had a relatively large memory and high processing speed that could accommodate the number of calculations required to illustrate a rudimentary design. The second development is associated with the mathematical modules that enabled geometrical calculations to become more usable (Peddie, 2013). Finally, the development of user-friendly pointer devices allowed the designers to easily manipulate what they saw on the monitor for the first time. Collectively these three advances could be regarded as creating the environment for the first CAD system.

In the early 1960s, the first two design-focused software systems to take advantage of these developments were Ivan Sutherland's *Sketchpad* and Jack Gilmore's *Electronic Drafting Machine (EDM)* (Weisberg, 2008). *Sketchpad*, the better known of the two, used a light pen and a set of buttons to facilitate interaction between the computer and a user (Sutherland, 1963). The light pen (initially a light gun) was a device with a sensor that detected the lights on the Cathode Ray Tube (CRT) monitor as the monitor refreshed the screen, allowing the software to pinpoint the exact position of the light pen relative to the screen. A button on the light pen coupled with others on a panel in front of the user, were used as commands. The pen's button, like that of a contemporary computer mouse, indicated the beginning and end of operations or constraints. The buttons on the computer panel indicated the commands themselves, such as for drawing a straight line, freehand drawing or drawing a circle. A significant feature of *Sketchpad* was its vector-based approach to geometry. Drawn objects were stored with their vector data and could be manipulated (for example, scaled or moved) by changing their constraints (Sutherland, 1963).

The other early CAD software, *EDM*, was initially developed by Gilmore, an MIT graduate like Sutherland, who also used TX series computers. He was employed by the defence contractor Itek Corporation, where he developed *EDM*. His software was similar to *Sketchpad* in that it used a light pen, a push-button interface and featured an object-oriented approach to the management of drawings. However, *EDM* ran on the smaller and more portable computer systems of the Programmed Data Processor (PDP) series, which were marketed to industries. In contrast, *Sketchpad* ran on the larger, more powerful TX-2, which was only at MIT and only used for research purposes. As a result of these factors, *EDM* was one of the first CAD tools to be marketed and used by design companies in the engineering and resources industries (Weisberg, 2008).

In the latter half of the 1960s, several companies and groups were formed by MIT graduates who had a core interest in developing CAD tools and improving software platforms. One of the early advances was the introduction of 3D graphics, initially in the form of orthographic perspectives in *Sketchpad III* (Johnson, 1963), and another was integration with printing hardware. Further developments of CAD tools, especially in the 1970s, improved 3D modelling capabilities (including the development of early NURBS surfaces), expanded their mathematical capacities to represent complex shapes such as Bezier curves, and integrated CAD tools with production processes. The availability of portable and personal computers, and introduction of operating systems based on multiple programming languages, were the catalysts for renewed interest in CAD development in the 1980s. These advances provided the opportunity for smaller design firms, especially in the urban planning and building construction fields, to use CAD tools.

There were, however, two major challenges facing the CAD pioneers of this era. Firstly, the hardware was prohibitively expensive. Even aviation and automotive companies could not necessarily afford systems capable of running CAD software effectively. Secondly, software authoring was a challenge because there was no standardised operating system, and each computer model used its own exclusive platform. Furthermore, each operating system was intricately tied to its underlying programming language, thereby complicating communication between devices and CAD developers.

Despite these challenges, researchers in the 1960s established several fundamental concepts in CAD which are still in use today. Firstly, they broke what was known as the "language barrier" between the user and the computer, which eliminated the need for users to compile their thinking into a machine language. CAD systems like *Sketchpad* eliminated this barrier by providing a digital analogue of the conventional pen and paper used by designers. The first of these was the light pen or pointer, then the drawing tablet, the trackball (which was less useful for designers), and the mouse (including light pen and tablet variations) which is to date the most common form of interaction between the designer and computer. Finally, CAD developers introduced object-oriented drawing, a system that completely changed the way designers interacted with a drawing. As Sutherland (1963 p. 18) observes, a computer cannot store a "drawing" in the sense of "a trail of carbon left on a piece of paper", but it could store the information that defined the drawing and its creation, which can be easily manipulated.

2.1.2. Evolution of digital modelling devices

The digital pointer or light pen effectively simulated the experience of designers using manual drawing tools (pens, pencils and rapido graphs), thereby creating the first combined hardware and software platform to support design (Prince, 1966). The ease of mathematical and geometrical processing enabled by computers also removed the necessity for other manual drawing tools, such as rulers, set-squares, compasses and templates. Furthermore, graphical advances in computing also provided more accurate and detailed drawing elements (including colour, texture and line weight), diminishing the need for equivalent manual tools.

A few years after Sutherland developed the light pen, Davis and Ellis (1964) invented the capacitive tablet and stylus. Named the *Rand tablet*, it became the progenitor of the modern drawing tablet. The tablet was both more accurate and faster than the light pen because it was not dependent on the screen's low resolution and slow update frequency (Weisberg, 2008). The tablet did however, introduce a level of dislocation into the designing experience, because designers could not see their own hands on the screen when drawing, or lines on the tablet where their hands were making them. The next major advancement in input devices, the mechanical computer "mouse", was developed in the 1970s (Englebert, 1970). The earliest 'mice' called "ball trackers" or "trackballs" required the user to roll the ball on top of a fixed device, instead of moving the device to roll the ball underneath it. The first computer mouse, however, offered a revolution in intuitive interaction for computer users. It was faster, more affordable, portable and relatively more ergonomic than the earlier pointer devices. With the rise of standardised, programmatically flexible operating systems in the 1990s, computer mice also became more compatible with other devices. Over the following decade the mouse became the preeminent input device, although additional features were added to provide CAD functionality. For example, extra buttons and wheels were used to facilitate additional commands, such as zooming, panning and rotating, in both 2D and 3D models. In contrast, tablets developed at a slower pace, the next major innovation being a pressure sensor that could offer an experience more reminiscent of conventional drawing.

The next generation of input devices was the touch sensitive display or "touchscreen". While this technology was initially costly, and mainly used in smaller devices, it soon found its way into larger screens suitable for CAD. The touchscreen initially had some of the properties of the early light pen, although it was faster, suitable for LCD screens and, most importantly did not require a person to hold a pen, as the screens were sensitive to fingertips. Touchscreens were soon developed that could detect multiple contact points ("multi-touch"), allowing a combination of positions, movements and taps ("gestures") which could be associated with specific commands. While current touchscreen technology lacks the practical accuracy of the mouse, its gestural capacity exceeds that of the mouse, because operating systems are usually not designed to capture the actions of multiple mice simultaneously.

These advancements in human-computer interface design were, however, limited to a 2D drawing interface. Even if the objects are depicted on screen as 3D, the hand and eye movements are bound to a 2D plane, which is either the flat screen, trackpad or mousepad. However, the idea of a 3D interface has long been postulated and explored in academia and industry (Seifi et al., 2019). This concept, grouped under the broad heading "haptic interface" (Briggs and Srinivasan, 2002), is generally realised by the use of "feedback" devices attached to hands. These devices capture, or simulate, the movements and gestures of the hands and fingers. Examples include sensitive tensile or mechanical arms with various 3D moves, wearable mechanical glove-like devices that sense the movement of joints and gloves with visual markers on them detected by an external sensor to help reproduce the gestures of hands. Sensors or markers may also be attached to an additional tool (like a paint brush or carving knife) to simulate their application. The online database *Haptipedia* contains more than a hundred patents, models and examples of haptic devices proposed or produced since 1992 (Yang et al., 2017).

Most haptic devices are used in two broad contexts. The first is through Augmented Reality (AR), where the haptic device provides a "tangible user interface" (TUI) (Shaer and Hornecker, 2009). AR works by overlaying an unrealised design onto a vision of the real world; the design can then be manipulated using the TUI or other detectable devices. The second approach is through Virtual Reality (VR), where it is the user themselves who is immersed into the virtual space of the digital model. In a simple VR variant, motion capture devices collect data for the system to simulate hands on the screen to manipulate the designed object (Chamaret et al., 2010). In a more advanced version, immersive VR, the designer wears a VR headset that allows them to feel they are physically present within the digital model's environment, and the designer may even be capable of walking around the digital model (by walking in the real world). An example of this variant is *VirtualBrick* by Arora et al. (2019).

Despite the promise and appeal of haptic devices, they are not yet widely used within the design industry. While some major software systems provide haptic solutions for reviewing models – for example, walking through an architectural design in Autodesk Revit – they do not provide a standardised and fully integrated haptic approach to design itself.

2.1.3. Evolution of CAD software and hardware

The history of CAD development in the 20th century could well be described as the shift from "computer aided drafting" to "computer aided designing." In order to understand how this shift occurred, it is important to trace the origins of CAD as a communicative medium, and then the evolution of its inputs and outputs.

The earliest description of *Sketchpad* (Sutherland, 1963) defined it as a "man-machine graphical communication system". Accordingly, the motivation for creating such a system was to facilitate interaction between humans and machines, which had previously been limited to key punching cards or slow typing of command strings. This desire to create an effective human-computer communication system was a driving force behind many of the early CAD developments in software and hardware (Ross, 1960). The drawing tablet, the mouse (Englebert, 1970) and precursors of video graphic adaptors like the *IBM 2550* (Weisberg, 2008), were all intended to improve this mode of communication. Software interfaces that minimised the designer's need to perform numerical or geometrical analysis before drawing were also created. In earlier CAD tools, this was achieved using simple drawing shapes as well as rudimentary modification functions, such as copying or erasing. For example, in *Sketchpad* a circle could only be drawn from a centre-point using a radius, whereas by the 1980s the earliest versions of *AutoCAD* included two additional methods for drawing a circle (Hall, 1986).

Importantly, the desire to achieve seamless communication between the designer and the computer included a consideration of both inputs and outputs. For the inputs, the focus was mostly on minimising the designers' efforts to prepare for and transfer their ideas to the CAD software. That led to the inclusion of various features in CAD applications which also allowed the emergence of related concepts such as Building Information Modelling (BIM) and Computer-aided Manufacturing (CAM).

The earliest CAD systems like *Sketchpad*, allowed only three types of input: insertion (drawing), modification and deletion. Insertion, much like its manual counterpart, allowed for the creation of shapes or symbols using numerical or positional constraints. Thus, rather than manually moving a pen on paper to create a circle (by following a circular stencil or the arc of a compass), in *Sketchpad* a circle was created by identifying a centre point and a point on its edge. The second type of input, modification, allowed for the variables that govern inserted shapes to be changed. In hand drawing the designer must physically erase part or all of a shape to modify it (or else draw over the top of it, rendering the original illegible); for example once a circle is drawn on paper the only way its radius can change is if the circle is erased and a new one produced. In contrast, the CAD modification function allowed some of the variables of an object to be modified because they were stored using those variables, not as pixels on a screen. This ability to modify was not limited to design or geometry, but could be used for most operations where data was stored as mathematical symbols. Later, the ability to "undo" and "redo" was also added to many software products, including CAD. Just as an object could be inserted or modified, so too could it be deleted. *Sketchpad* had only these three input types, and two geometrical entities (line and circle), but within a few years most CAD software included tools to create 2D arcs, ovals, rectangles, polylines and customisable curves. 3D entities, like surfaces, cubes, spheres, and cylinders were also soon included, creating new primitives from which to construct a design. The modification function also evolved to include Boolean operations, finishing (fillet, chamfer, bevel, among others) and extensions (such as extruding or offsets). Delete, while remaining relatively straightforward in principle, could also be controlled to determine which operations or objects would be subject to this function.

The next stage in the development of inputs for CAD was the inclusion of standard or replicable components. Mass-production, standardisation and drawing conventions have led to certain units (produced or drawn) being used multiple times in one or across multiple designs. In manual drafting, each draftsperson or designer would have to draw a scaled representation of the repeated unit each time, or use a stencil, or they could represent the unit using abstract symbols. However, CAD provided the means to design and replicate items as groups or blocks, usually with the ability to automatically update the replicated instances by modifying the original unit. Furthermore, once a repeated part was changed, associated parts of a design could be automatically updated to reflect the change and its implications. For example, the scale, orientation or position of different parts in a machine are all connected to each other, so that by changing a central feature, the rest of the machine will behave as if it were actually physically connected in this way.

An output, in a computational sense, is a visual representation of the designed product in a computational medium, typically on a screen. It should not be mistaken for the actual output of the design process, that is, the resultant product or prototype. For improving CAD outputs, developers initially focused on improving the quality of the illustrated design elements and shortening the time for this illustration to occur, to the point that the input and output are almost simultaneous. However, in the 20th century this objective was dependent on processing speed and video technology. For example, while Bezier curves had been used in CAD since the 1960s, it was only in the late 1980s that CAD tools began to render them in real-time. Furthermore, the difficulty of rendering increased in line with the level of realism required in the 3D representation. Realistic representation often involved 3D non-isometric projections with materials and lighting. Reflective materials or transparency would make rendering even more time-consuming. The main technological development to solve this problem was the video graphic adapter, or "graphics card" which, in combination with software interfaces (like *OpenGL* or *DirectX*) allowed most of the rendering be done by optimised hardware.

Nevertheless, despite technical advances, fast realistic rendering is still not an option for large projects and many computer systems. More importantly, it is not always useful while designing. While CAD systems usually have options for such outputs, they also feature a set of illustration conventions to provide the designer with strong visual control over the design. These usually include several view directions (often depicted as the six sides of a cube), different types and foci of perspective, various types of rendering (such as wireframe, hidden, shaded, etc.) and layers. In addition, they offer viewing commands such as zooming, panning, rotating or sliding the view.

Most early CAD systems combined software and hardware and were often variations of generic computer systems. Due to their cost, CAD usage was often limited to large engineering design firms, notably in the automotive and aviation industries. Indeed, one major contemporary CAD developer, *Dessault Systèmes* (DS), started as a subsidiary of a French aviation company. However, since then CAD has undergone five development stages, each driven by advances in microprocessing, rendering, scripting, BIM and dynamic data linkage respectively (Lai, 2017).

The first stage in the evolution of CAD can be traced to the introduction of microprocessors in the 1970s, which led to the production of smaller and more affordable computer devices. In the 1980s, devices like the *IBM 5150* and the *Macintosh* were already accessible to many people. The standardised operating systems and interfaces of these computers created the opportunity for software developers to distance themselves from hardware design. In this environment several CAD packages were developed for smaller-scale firms or individual designers. Two of the most famous of these are *AutoCAD* by Autodesk and *CATIA* by DS. Both packages were intended for engineering design, with *AutoCAD*'s early versions limited to 2D drafting while *CATIA* featured some additional 3D manipulation. These packages offered many of the standard tools and capabilities of the CAD software in use today.

The second stage in CAD evolution was associated with the introduction of 3D graphical rendering. While 3D capabilities had already been part of some CAD software in the 1970s, they were limited to wire frame models. In the 1980s, 3D modelling tools were mostly associated with the movie industry rather than engineering design. *SoftImage* and *Preview* were some of the tools that were created for animation or visual effects. In 1986, Autodesk released *AutoFlix* and *AutoShade*, its first rendering extensions for *AutoCAD*, and a few years later it released *3D Studio* software for 3D modelling, rendering and animation. By the 1990s and 2000s, CAD packages could be classified into two broad categories in terms of their approaches to graphical representation and rendering. The first was focused on the engineering aspects of design. In this category, software packages like *AutoCAD*, *SolidWorks*, *CATIA* and *Microstation* supported fast numerical, textual and geometrical operations by designers, and quick, if rudimentary, visual feedback in return. The second category included packages like *3D Studio*, *Maya* and *Blender3D*, which were focused on visualisation and modelling, providing powerful freehand manipulation, rendering and animation tools. In the last decade, the difference between these two has become blurred, with "plug-ins" and robust file export forms allowing engineering designs to be rendered realistically, or animated models analysed mathematically. In both cases, advances in software and hardware development supported more efficient visual representations of the design either via their built-in functions or extended add-ins.

The third stage of CAD evolution is associated with the development of scripting capacity in CAD software, allowing the designer to customise the interface or commands. Prior to that, the controlled or standardised nature of CAD systems restricted designers from taking a more comprehensive approach to customising their workflow operations. In the third stage, scripts provided the opportunity to automate aspects of design or drafting. An early example was *AutoLISP*, a LISP-based programming environment included in *AutoCAD* in 1986. In the 1990s, software advancements, such as the introduction of the Windows operating system and its run-time libraries, as well as the increased use of Object-oriented Programming (OOP) languages such as Visual Basic, made the incorporation of programming features within software packages more viable. With OOP, the designer-programmer was able to easily access and manipulate CAD objects, their properties and hierarchies using scripts. *AutoCAD* included two OOP interfaces, Microsoft's standard Visual Basic for Applications (*VBA*) and *ObjectArx*. *VBA* was also used in other CAD packages like *CATIA*, *SolidWorks* and *3D Studio*. Later, several CAD packages featured scripting environments, such as *MaxScript* for *3D Studio* and *Grasshopper* for *Rhino*. Furthermore, by using shared libraries (such as DLLs in Windows) many CAD applications allowed their design and environmental content to be accessed from general programming environments regardless of their language.

The increase in interest in object-oriented processes was not, however, a product of OOP's inclusion in CAD. It was driven by a growing desire to include semantics in CAD geometry. In the early 1990s, information management emerged as an important domain in industries and businesses (Venkatraman, 1994). This change, combined with heightened awareness of a product's life cycle, resulted in a new focus on design and product management. In design computation, this concept was materialised through BIM, which sought to manage the project's design and data digitally during its life cycle (Muñoz-La Rivera et al., 2019). Although the term "building" has its formal origin in architectural design (Ruffle, 1986), its use has been expanded to all domains of engineering design.

BIM initiated the fourth stage of CAD development. The first contribution of BIM to CAD was to distinguish the definition of designed entities from their geometrical properties. The other major contribution of BIM was to facilitate linking of entities or objects and their properties. For example, in *AutoCAD*, to design either a wall or a perforated plate a closed polyline is drawn and then extruded into the desired elevation, and then openings or holes are made by removing other solids. The main differences between these solids, as far as the software is concerned, is their geometry. However, in BIM-enabled CAD software, like *Autodesk Revit*, the user designs a wall as one, and then select predefined openings and they are recorded as such. The designer can add non-geometrical information, such as material or structural properties, and digitally link this with other information about the wall, like the services, ducts or conduits passing through it.

The linking of information to objects and their relationships facilitated the fifth stage of CAD development. Such linkages can be either static or dynamic. In the former case, the software only retains fixed information about an object, say, the position and width of an opening relative to the wall. In the latter case, dynamic linkages, changes in one object will automatically change those it is connected to. For example, a change in either the wall or the opening in it will automatically update the other. This dynamic linkage is often associated with parametric design (see the next section). Since the 2000s, CAD software has increasingly incorporated parametric features, which are either realised by using pre-defined associative properties (for example, how walls and openings can interact) or by way of explicit programmatic abilities incorporated in CAD/BIM packages. In the latter case, some CAD packages now offer specialised scripting environments, such as *Dynamo* in *Revit* or *Grasshopper* in *Rhino*, to facilitate parametric designs.

2.2. Parametric and generative design

Some of the most significant recent developments in the evolution of

CAD are associated with generative and parametric design (Ostwald, 2017). Generative design is a broad category that covers automated or rule-based creative processes, whereas parametric design is more often understood as a specific type of generative design. Both involve the move away from designing with geometry to designing with algorithms. This section commences with a brief overview of generative design categories or types before focusing on parametric design.

2.2.1. Key concepts and characteristics of generative design

The term *generative* first appeared in design and visual art in 1965, at an exhibition of computer art in Stuttgart called "Generative Computergraphik". Since then, it has gradually evolved from being a description of design mechanics (as computer-generated design) to a more specific reference to design paradigms capable of supporting design thinking (Eckert et al., 1999). The origins of generative design are often traced to the 1950s and 1960s enthusiasm for natural phenomena. For example, evolutionary systems and genetic algorithms were inspired by the processes of biological evolution (Holland, 1975), such as cellular automata which were modelled on biological growth (von Neumann, 1951) and design grammars which had synergies with formulating natural languages (Stiny and Gips, 1972). More recently, generative design has been focused on both shape-oriented and other functionally-defined algorithms. Generative design has also gradually become more common in design-related disciplines, such as architecture, product design, aviation and automotive design.

Many scholars have offered definitions of generative design. For example, Agkathidis (2015) describes generative design as a design method where the generation of form is based on rules or algorithms. Lazzeroni et al. (2012) define generative design as a cyclical process based on a simple abstracted idea that is applied to a rule or algorithm. Most of the definitions of generative design identify algorithmic operations as a main feature. An algorithm is a set or sequence of mathematical rules or directions that can be applied to produce a result, response or solution. In the context of computational design, an algorithm is a set of rules and processes used for processing information to create a solution to a design problem. In summary, generative design is a collective term for a number of computational methods or algorithmic approaches used to automatically create or evolve design solutions. In contemporary use, such methods use algorithmic procedures to produce large numbers of design alternatives, from which the most suitable design may be selected (Herr and Kvan, 2007). This definition has been systematically used since the 1990s in the computational design domains, although many generative design systems were developed prior to this time.

Generative design methods typically have three components: *algorithm*, *ideation* (for generating design alternatives) and *computation*.

In most generative methods, *algorithms* are realised as a set of generative rules (Boden and Edmonds, 2007; Oxman, 1990), *build commands* (Hornby and Pollack, 2001) or *transformation rules* (Gero and Kazakov, 1996). Generative rules are responsible for changing the form in all or part of a design into a new one. Rules differ by the mechanisms of generation they adopt. One example of this is a shape grammar, which was originally proposed by Stiny and Gips (1972). As Stiny (1976) states, a "shape grammar is defined over a set of shapes, and maps into shapes to generate a shape language" (p. 191). The generative mechanism in a shape rule can be used for either replacement or modification (Knight, 2003). Replacement is the substitution of the design or its part(s) with another. It can also include addition (replacing a void with an element) and subtraction (replacing an element with a void). Conversely, a modification changes the scale, orientation and direction, or other numerical properties of a design. Through the application of modification rules, design generation can be visualised as a sequential process (Herr and Kvan, 2007). This process selects a combination of rules to advance the design generation process in a structured manner, from an initial state to the final outcome. Design alternatives emerge when different combinations of rules are selected and applied. These sequences are not always explicitly defined, and may occur interactively during their application. For example, in shape grammars, rule sequencing is made possible in real-time by the recursive matching of existing shapes from the design with the Left-Hand Side (LHS) shapes of the rule set (Maher, 1990). Another more advanced type of sequencing is through cyclic reproduction of design (Eckert et al., 1999). In this case, rules that have directed the generation of the previous stage will undergo further rounds of application, but only on the design with the most potential alternatives identified during the process. Such a process is intrinsic to "evolutionary design".

Ideation refers to the process of generating alternatives that satisfy design requirements (sometimes called "design divergence"). Usually, this ideation process is enabled by the selection of multiple rules and their sequencing for application during design generation. For example, in shape grammars, the decidability of LHS rules contributes to the divergence of the design results (Knight, 2003). Other factors can also play a role in enabling design divergence. For example, in parametric shape grammars the range of parameters may allow multiple definitions of design by adjusting their values (Knight, 2003).

The last set of characteristics, *computation*, refers to the fact that generative design has a formal structure comprising visual or mathematical properties applied in a systematic way. Due to the increasing complexity and intensity of the design generation process, research has extended the notion of "computation" to the use of computers as well (Boden and Edmonds, 2007). In some domains, for example, "computer art" and "generative art" have been used interchangeably deriving from this context.

2.2.2. Major generative design systems

Multiple generative design systems have been developed since the 1960s, many of which share common origins or features which are useful for categorising them. For example, Fischer and Herr (2000) characterise generative systems into four types: *emergent systems* (for example, cellular automata), *generative grammars*, *algorithmic generation* (including parametric design) and *algorithmic reproduction* (genetic algorithms). While this categorisation assists researchers to distinguish common features, the categories are not precise. For example, the phrase "algorithmic generation" can arguably be applicable in all generative systems. The term "emergent systems" appears to refer to support for design emergence, although it is also applicable to a number of systems. An alternative approach to categorising generative systems simply acknowledges that generative systems can have both similarities and differences between each other (Singh and Gu, 2011). This section uses the former four categories for the review, but considers both their differences and similarities, and therefore does not treat the categories as mutually exclusive. The four categories are generative grammars, evolutionary systems, self-organised emergent systems and associative generation.

Generative grammars

Generative grammars use transformational rules that can be applied recursively to develop an object or shape. The concept of "grammar" originated in linguistics but has evolved into different forms for unique purposes. For example, visual grammars, such as shape grammars, were inspired by Chomsky's idea of generative grammars (Stiny and Gips, 1972). Graph grammars were initially developed in computer science and are especially suitable for computer implementation (Chakrabarti et al., 2011). *L-systems* were based on another linguistic concept, string grammars, which were developed in the 1950s (Singh and Gu, 2011). In recent years, research has combined these approaches to enhance generative capabilities. For example, combinations of shape and graph grammars (Grasl and Economou, 2013; Lee et al., 2016) have been used for addressing both spatio-visual and syntatic issues in architecture. Generative grammars are amongst the earliest generative design systems to be used in architectural design. One of the most famous of these was the shape grammar of Palladian villas by Stiny and Mitchell (1978). On the other hand, graph grammars and L-systems have only been applied to architectural design since the 2000s (Parish and Müller, 2001).

Generative grammars are sequential due to the nature of their rule application process. The sequencing of rules for application is determined by matching different existing elements of the design against LHS shapes in the rule set for their selection (Stiny and Gips, 1972). The divergent generation of design alternatives could be fulfilled by different possibilities in rule selection and application processes. While all grammars are computational, only *graph grammars* and *L-systems* have been substantially implemented using computers. One reason for this is that during shape detection, shape grammars allow designers to freely decompose and recompose shapes. This feature is called shape *emergence*, which regards them as not being predefined but emergent. To appropriately handle shape emergence has been difficult in computer implementations because the finite representations of a design are restricted by the computer memory (Krish, 2011; Tching et al., 2016), which therefore becomes an ongoing challenge for generative grammars.

Evolutionary systems

Generative design systems and genetic algorithms emerged in the early stages of computational design in the late 1960s (Holland, 1975), although their explicit application in architecture began in the 1990s (Jo and Gero, 1998). As the title suggests, evolutionary systems in design were inspired by evolutionary biology, especially the simplified notion of "survival of the fittest" (Holland, 1975). An evolutionary system is one that defines a recursive process of design reproduction in which each object state (conceptualised as an "organism") reproduces a divergent state (or "offspring"). Only the "fittest" amongst these offsprings is allowed to trigger further reiteration or reproduction (Fasoulaki, 2007). In an evolutionary system, the "fitness" of a generated design instance is evaluated using criteria to select the most appropriate ones to continue the generation and selection cycle until a satisfactory outcome is achieved. Due to the analogical origins of this idea in evolutionary biology, these generative design systems are also described as genetic algorithms (GA). It is, however, critical to remember that there is nothing natural or organic about a genetic algorithm or an evolutionary system. The words, "genetic" and "evolutionary" simply signal that a computational process is analogous to, or superficially reminiscent of one in nature. In an evolutionary algorithm the designer must set the rules (they are not predefined) using some other logic system, such as shape grammars or L-systems. When sequentially applied to a cycle of generation and selection, these generative algorithms become evolutionary algorithms.

One of the benefits of evolutionary systems in design computing is that their outcomes are usually sizable in terms of both depth (number of stages in design generation) and breadth (number of alternatives per stage). In most evolutionary design systems, the computer is used to assist in managing the generation and selection cycle. However, in cases where the selection criteria are more subjective (aesthetics, for example), designers choose their own interventions (Ostwald, 2004, 2010). The challenge, however, is in formulating explicit fitness criteria, because without them evolutionary algorithms are just facile form-making machines.

Emergent and self-organised systems

The concept of "emergence" refers to the explicit outcomes of implicit organisations (Gero, 1996). Therefore, emergent systems are those whose outcomes emerge out of self-organised components. A common approach to achieving emergence in computational design is to assemble a collection of self-organising agents, which shape the final form by interacting with each other and their surroundings. Cellular automata and swarm intelligence are examples of emergent and self-organised systems (Singh and Gu, 2011).

Cellular automata systems, which were inspired by an interest in simulating biological growth (von Neumann, 1951), were formalised in the 1980s (Wolfram, 1986) and have attracted the attention of designers since then. In practice, a cellular automata system is typically founded on a structured geometry, most often a grid, whose cells can have different properties or states (Sarkar, 2000). The rules or sets of rules associated with the cells direct the alteration of states in response to changes in neighbouring cells or time intervals (Batty, 1997a). As time passes, the cells automatically change their properties or states and the final state, sometimes called the emergent condition or emergent design, is the outcome. Due to this automation process, the term cellular automata is often used to describe such generative design systems.

The abstract nature of the cells allows the generative rules to combine both replacement and modification functions as generative mechanisms. The automation process is also sequential, as state alteration proceeds and spreads cell by cell. Design divergence in cellular automata can be achieved either by recording the collective states of cells at different times, or enabling parameters and variations that can affect the automation rules (Batty, 1997b).

Swarm intelligence refers to systems where the collective behaviours of simple individual agents work collaboratively to produce an outcome (Blum and Li, 2008). Like the cells in cellular automata, agents in swarm intelligence behave collectively in accordance with their inherent rules, as well as their interactions with each other and their surroundings. However, unlike cellular automata, the design outcomes of swarm intelligence are not bound to the defined field or grid (Singh and Gu, 2011). Furthermore, their behaviours can be more complex than those of the alteration of states in cellular automata. Swarm intelligence was inspired by the study and simulation of insect swarms and has been extended to consider other animal and human behaviours (Garnier et al., 2007). The agents in swarm intelligence usually operate on time intervals, which define the sequential characteristics of design generation. The divergence of the design outcomes may be realised by various means, such as varying (or even randomising) the numbers and types of agents, their surroundings and time intervals.

Associative generation

In an associative generative system, the design and its alternatives are generated by first defining the relationships between different components that make up a design and then by assigning values to both. Unlike generative grammars, in associative generative systems design components are not explicitly transformed, but any changes made to their properties will subsequently lead to the generation of new instances by adjusting other properties due to the defined association. Parametric design, which is the focus of the remainder of this section, is the most well-known associative generative system.

The essence of parametric design is that components and their relationships are defined by parameters that can be manipulated directly, rather than by simply manipulating components. This concept has also been adopted to enhance other generative design systems, including those such as parametric shape grammars. In theory, parameters and variables can be of any type or value, including those related to visual qualities, although in practice they are usually numerical or algorithmic. Divergence may be supported by either adjusting the parameters and variables, or modifying and redefining the association between them. This allows for the efficient production of a large number of design alternatives, which can then be explored and reviewed to finalise the design generation.

2.2.3. Key concepts in parametric design

Parametric design grew out of the larger field of generative design and has since evolved into one of the most widely applied computational design approaches in both practice and education. This section commences with an introduction to fundamental concepts in parametric design. The next section then describes historical background and development and gives examples of applications and implementations.

This section defines three key concepts: parametric design, parametric variations and parametric modelling.

Parametric design

A parameter is a value or measurement of a variable that can be altered or changed. Each object in a parametric system may have certain rules embedded in them, and when one parameter changes other parameters will adapt automatically (Ostwald, 2012). By changing various parameters, particular instances can be created from a potentially infinite range of possibilities (Kolarevic, 2003). Parametric design focuses on the representation and control of the relations between objects in a computational design model. Using parametric design tools, designers can create rules to fulfil aesthetic, functional or performance-related requirements of a design. Through this process, parametric design supports the creation, management and organisation of complex designs (Woodbury et al., 2007).

Parametric design systems can be understood as having the following four characteristics. First, they are structured around design parameters. Eastman (2008) suggests that in parametric design key variables are often defined by parameters. While most parameters are related to geometrical modelling, others can be connected to functional or performance-related requirements. Second, they structure the relations between design variables. As Cárdenas (2007) notes, parametric design establishes the relations between modelling components defined by constraints, while Abdelsalam (2009) proposes that the relations are maintained by variations. In most cases, both constraints and variations in combination define the relations between the modelling components. However, a parametric system may also become "over-constrained" if there is no effective control over the defined relations (Burry, 2003). Third, in a parametric system, rule-based algorithms ensure that the design process is flexible but controllable (Schnabel, 2007). As Abdelsalam (2009) argues, by making rules, designers can produce variations that result in "fully organised controllable building forms" (p. 299). Fourth, parametric design provides a formal process from which multiple design solutions can be developed simultaneously and efficiently. Both Hernandez (2006) and Karle and Kelly (2011) emphasise that to develop parallel ideas is one of the main advantages of parametric design.

In summary, parametric design is a dynamic, rule-based process controlled by variations and constraints, in which multiple design solutions can be developed in parallel. It is especially effective in generating complex forms and optimising multiple design solutions. As a result, parametric design is usually utilised in complex building form generation and fabrication, structural and energy optimisation, and other design iteration and prototyping tasks. Parametric design differs from traditional CAD approaches because of its use of rule-based algorithms.

Parametric variations

As the previous section noted, in a parametric design system variations are controlled by changing the values of parameters and constraints, without necessarily altering the original structure of the model (see Figure 2.1). Variations can be single or multiple; independent from or interrelated to each other. Importantly, Karle and Kelly (2011) argue that parametric design does not force designers to generate an ideal design solution, but rather to ask the right question, which can then be answered with multiple solutions. As a result, the selection process of the generated variations is an important step in parametric design. Prior to selecting the appropriate variation(s), a typical workflow in parametric design also involves identifying the design problem(s) and developing a series of rule-sets with associative design variables that can support a flexible and dynamic design process, enabling the emergence of variations.

Design constraints play a significant role in directing the parametric process. They can be used by designers to describe and generate a range of variations, and to define and control the unique characteristics of each. Constraints set limits on parameters and control the possibilities for the range of variations. In parametric design, constraint satisfaction is important in the decision-making process, by connecting individual factors with the overall design outcomes. There are two basic forms of constraints: geometric and dimensional (Monedero, 2000). Geometric constraints are properties that control how geometrical entities relate to each other. Dimensional constraints are properties that can be assigned a singular value that fixes its behaviour until it is changed or removed. An optimal parametric design process often requires the computational model to be well balanced, which is neither under-constrained nor over-constrained (Monedero, 2000).

Parametric modelling

Parametric modelling refers to the techniques used to "create design spaces and manage geometric dependencies within a model" (Gane and Haymaker, 2009, p. 81). It provides a formal descriptive and generative design framework through parameters and their associative relations by which designers are able to change the input values to generate and optimise design and variations. The most significant advantage of parametric modelling is that it allows changes to be made to parameters at any stage of the design process (Monedero, 2000). Different parametric modelling techniques have been developed for visual purposes (for example, form-finding), as well as for other functional or performance-related purposes. Some typical parametric modelling techniques for form-finding include repetition and subdivision (Figure 2.2) along with tiling, recursion and weaving.

Research further synthesises and formalises essential parametric modelling techniques as "design patterns" (Woodbury et al., 2007). Parametric design patterns are reusable tools that address particular

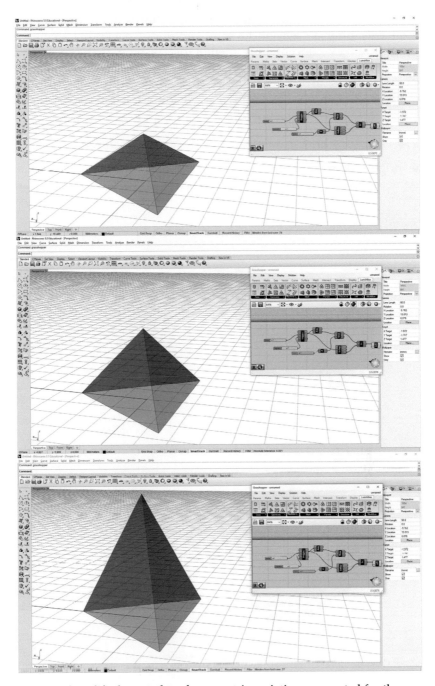

Figure 2.1. Simplified examples of parametric variations generated for the same geometric primitive (the cone).

problems. The advantage of these general patterns in parametric design is that they offer a "way to identify successful general strategies that exemplify a key concept in a memorable fashion that can easily be taught" (Woodbury et al., 2007, p. 229). By using and learning these general patterns, architects and students alike may be able to master parametric design more efficiently and skilfully through abstraction and standardisation (Woodbury, 2010).

2.2.4. History and application of parametric design

The origins of parametric design are typically traced to 19th century parametric equations in mathematics, with the first use of the phrase

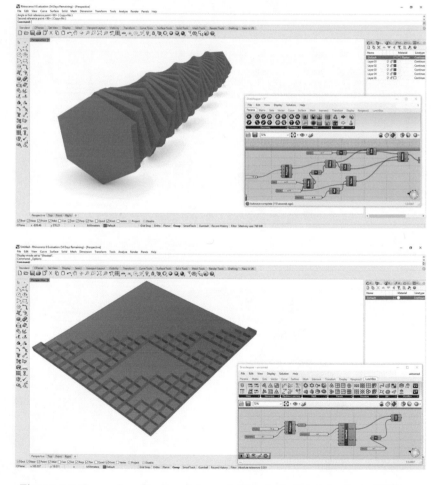

Figure 2.2. Examples of parametric modelling techniques for form finding: repetition (top), and subdivision (bottom).

"parametric architecture" being found in the work of Italian architect Luigi Moretti in the 1940s (Frazer, 2016). There are examples, however, of simple numerical design parameters in the historic architectural treatises of ancient Rome (Vitruvius) and the Renaissance (Serlio), and as such, it is not an entirely new concept in design. For example, two decades before Moretti, an early practical example of variation control can be seen in architect Antonio Gaudi's design for the Church of Colònia Güell (1908–1917) in Barcelona. For this design Gaudi created an inverted model of the ceiling shape of the church (hung upside down), that could be altered by adjusting the position of weights or lengths of string (representing, structural loads and elements respectively). The whole structure of the model responded to minor changes in each element, demonstrating a fundamental concept of parametric design. Gaudi used the same principle for his design of the Sagrada Familia in Barcelona (Burry, 2003).

A similar variation control concept inspired Hillyard and Braid (1978) to propose a system for computer-aided mechanical design that allowed the specification of geometric constraints within a certain range. A few years later, Light and Gossard (1982) developed and presented "variation geometry" or "variation design". Their work provided geometrical representations with enhanced mathematical and geometrical modelling tools to support the design process. Parametric modelling tools of this type soon began to be utilised in aerospace, naval architecture and product design industries. From the late 1980s, Frank Gehry, a leading figure in contemporary architecture, used *Computer Aided Three-dimensional Interactive Application (CATIA)* for the documentation of his building designs. *CATIA* was first developed in 1977 by Avions Dassault. Thereafter, it was used in aerospace, automotive, ship-building and other industries for its capacity to control and manipulate complex geometries as well as to support improved manufacturing accuracy. A component of this platform, a parametric package for modelling *NURBs* surfacing, began to be used in commercial architectural practices after this time (Dal Co, 1998). Although Gehry's designs of the era used parametric techniques predominantly in the documentation stage, his aesthetic style suggested new architectonic possibilities and triggered intellectual debates about the style and the role of the computer (Forster, 1998; Zellner, 1999; Ostwald, 2006, 2015). Without parametric tools, these dynamic open-ended, complex designs would have been difficult to achieve. Based on *CATIA 5.0,* Gehry Technologies (GT) later standardised this parametric approach into their *Digital Project (DP)* software to handle complex architectural designs. Today GT has expanded its services to a wide range of computational design areas. *DP* is now a powerful parametric software package that can effectively handle complex parametric as well as geometric associations, which make it ideal for large design projects.

From the mid-1990s, the number of architectural practices and academic organisations exploring or adopting parametric design has increased significantly. During that period, software such as *GenerativeComponents™ (GC)*, *Grasshopper* and *Processing* were developed and evolved rapidly. In 2005, *GenerativeComponents (GC)* was developed by Robert Aish for Bentley Systems. *GC* implements parametric concepts across the entire design project lifecycle, from the early conceptual phase to the final documentation. In addition, *GC* has been integrated with *Building Information Modelling (BIM)* as well as other analysis and simulation platforms for design evaluation and optimisation. This type of integration can potentially make parametric design more targeted and realistic, linking design conceptualisation with production, fabrication and construction. *Rhinoceros (Rhino)* is a stand-alone, NURBS-based 3D modelling tool developed by Robert McNeel. It has since been widely adopted in a range of domains including architecture, industrial design, jewellery design, automotive as well as marine design. *Grasshopper* is a rule-algorithm editor with a graphic interface that can be integrated into *Rhino* as a scripting plug-in. It is structured with specific definition files that link to the main parametric model in *Rhino*. *Grasshopper* is generally used as a generative tool rather than a modifier during the parametric design process. Compared to other parametric software, *Rhino + Grasshopper* has been much more widely adopted in both practice and education in recent years. This is at least partially because of its ease of operation as a visual programming tool (supported by the graphical rule-algorithm editor) and its relatively low cost.

In parametric design, scripting uses computer programming languages such as *Java* and *Visual Basic (VB)* to directly interpret and execute commands. It can be used to establish and control parameters and translate design intentions into codes that are easily identified by a computer. Consequently, designers with fluent scripting skills can define and control the rules for parametric design in a way that allows them more freedom than other tools. Common scripting languages include *Python*, *VB* and *Ruby*, whilst scripting tools include *Python Script*, *RhinoScript*, *Processing* and *CADscript*. There are also several design analysis tools that are often used together with parametric software. For example, *Ecotect* is used for analysing energy performance, while *ETABS* is used for structural analysis. These analysis tools are capable of exporting data into parametric software to direct design generation and optimisation.

By the year 2000, applications of parametric techniques in building design had grown in frequency and sophistication. The leading architectural practices using parametric design at the time included Foster+Partners, Zaha Hadid Architects, UNStudio, KPF, AAEmtech and SPAN. Designers were able to use different parametric design tools to produce and control free-form architecture. This was a major breakthrough for many architectural practices that was enabled by the adoption of parametric modelling. In parallel with this, in research and education parametric design also expanded at a great rate (Ostwald, 2017). Some architecture schools – AA in London, LAAC in Spain, Hyperbody in Delft University of Technology, MIT and Columbia University – soon developed reputations for parametric design, although within a few years almost every major architecture school was teaching or researching in this field. In parallel, multiple international computational design conferences – ACADIA, ASCAAD, CAADRIA, eCAADe, SiGraDi, and CAAD Futures – published research on parametric design.

At the 2008 Venice Biennial titled *Out There: Architecture Beyond Building*, the term *parametricism* was used to describe a combination of design concepts that "offer a new, complex order via the principles of differentiation and correlation" (Schumacher, 2009, p. 15). Parametricism was also presented as a design style or movement to replace Modernism (Schumacher, 2009). Despite such arguments, there is a clear difference between computational parametric design and the parametricist aesthetic. The former is a process or technique that can be used to create or optimise often quite conventional-looking, orthogonal buildings that do not fit Schumacher's (2009) definition. The latter organic, free-flowing, folded or curved buildings can, and mostly still are, produced without using parametric methods. An example of the former approach is the Soho Shang Du building in Beijing, which was designed using parametric tools and has a rather conventional appearance. Conversely, Federation Square in Melbourne is often cited as an example of generative design, but its design process was largely conventional, manual and arts-based.

Applications of parametric design

Parametric design tools can be used to capture and explore critical relations between design intentions and geometries. Designers interact with these tools using rule-algorithms to define and manipulate their relations as well as relations between different design elements. For the purposes of "form-finding", parametric design can be especially useful for facilitating the modelling process for complex geometries, and the integration of parametric tools in design can also enhance flexibility and control during the process (Fischer et al., 2003). To explore form-finding in parametric design, some studies have focused on geometrical modelling methods. For example, Hnizda (2009) suggested that there are at least two different geometrical modelling methods in parametric design – *object extraction* and *transformation* – each of which can be used to explore the relations between formal aesthetics and functional properties. Other researchers, such as Baerlecken et al. (2010), have explored these issues through problem definition in the early conceptual design stage. To a large extent, the latter form-finding approach through problem definition is dependent on variation settings, and through parametric variations they have explored functional and structural properties that result from requirements such as sun-shading and the aesthetics needs of the final design form.

When using parametric tools for structural design, Maher and Burry (2003) compared parametric structural analysis with traditional approaches in a cross-disciplinary collaboration between architects and structural engineers. The combination of structural analysis and parametric design not only enhanced the design process and outcome, but also provided a new platform for collaboration across disciplines. In a similar study linking architectural design to structural optimisation, Holzer et al. (2007) investigated geometry generation using structural analysis and optimisation, showing that parametric systems are effective for generating a variety of solutions for structural design. Most of these works have been focused on analysing the volume of structural materials. In contrast *ETABS* is a popular structural analysis tool whose data can be imported into parametric design software. For instance, Almusharaf and Elnimeiri (2010) studied structural performance in high-rise building design using *ETABS* in the *Grasshopper* parametric environment. A design scenario was presented in their study where instant feedback of structural performance can be provided during the parametric process to assist decision making and design generation.

One of the advantages of parametric design systems is that multiple analyses can be established and conducted in parallel within the same model to support more comprehensive optimisation. Besides structural analysis, building performance especially in terms of sustainability, is another important application of parametric design. For instance, "multi-parametric façade elements" were proposed and examined by Schlueter and Thesseling (2008). In their study, the performance analysis of the façade offers instant assessment (in terms of the solar gain in different façade forms) during the design process, so that designers can better improve energy performance while generating design forms. In another study, a formal framework that combines parametric modelling with a "performance-based design" paradigm was proposed (Bernal, 2011), aiming to standardise parametric applications for building performance, providing real-time feedback during the design process.

Other than form-finding, and with the continuing evolution and growing popularity of parametric design, researchers have begun to explore design collaboration using parametric systems, including cross-disciplinary and geographically asynchronous scenarios. Cross- and multi-disciplinary collaboration is considered to be especially beneficial in the early design stage, and the use of parametric design at that time can potentially assist teams in problem finding, which in turn directs the generation and selection of variations during the parametric process (Gane and Haymaker, 2009). In terms of collaboration between different locations, Burry and Holzer (2009) explored the potential for sharing parametric models in a version-control platform that could be used by design teams in different cities. Rajus et al. (2010) used the participant observation method to study 18 participants in different locations working with parametric design tools. In their study, participants were asked to perform different design tasks using *GenerativeComponents*. Their results show that a moderately controlled collaboration process can enhance team performance and user satisfaction in parametric design.

2.3. Presence and telepresence, virtual and real worlds

Most design processes involve visualising a product that does not exist yet, or in other words, is not present in the real world. The product only exists as information on paper, in computer models, or sometimes as scaled physical models. For designers it is important to visualise, how a design may look and feel for an inhabitant, or be held by a user. In addition to aesthetic considerations, the capacity to look around and inside a design also supports improvements in terms of functionality and safety. The desire to preview a design is also important for clients, approval authorities and consumers. Traditionally, designers used a plethora of media to visualise a design's properties and present them to different stakeholders. These include 2D perspective sketches or renderings on paper, orthographic drawings (plans and elevations), schematic models and "walk-throughs" or 360-degree "rotation clips" of a design. However, technological advances have provided new opportunities for designers, clients, and consumers to experience an as yet unrealised design without it being physically available (Schnabel, 2008). This opportunity has arisen because of advances in telepresence, Virtual Reality (VR), and Augmented Reality (AR) (Gu and Ostwald, 2012; Ostwald, 2012). This section introduces and discusses all three, along with their application to design.

2.3.1. Telepresence

Telepresence, or "distant presence", describes a capacity to be able to sense or interact with a real place or person, to a relatively realistic degree, while not being physically present there. The most basic type of telepresence is associated with audio-visual communication devices, which transfer information over a distance. More complex and recent forms include remote controlled machines like drones and robots which operate in conditions too dangerous or inaccessible for humans. Using these devices, the human operator may interact with the distant environment by actively manipulating it (for example, repairing or rescuing). Telepresence, *per se*, is not usually employed for product design because it interacts with completely real objects and environments which do not provide other options for essential design tasks such as ideation, evaluation, or prototyping. However, it may involve tasks aligned with production or construction. A common example is a surgical or invasive medical procedure using small devices which provide vision and access inside the body, with a minimal requirement to cut or open tissues. In architecture, it has become common for large multinational design firms to have telepresence studios in different locations (both offices and sites) around the world, and to constantly broadcast the sound, vision and computer screens from these locations to all of their other telepresence rooms for designing in teams in real-time across diverse locations.

The concept of telepresence enables and promotes distant collaborative design in architecture and design. However, designing in teams, also known as *collaborative design*, involves increased complexity and difficulty in the design process (Larsson, 2007). When collaborative design is conducted remotely, the complexity and challenges are increased. Digital tools that serve as shared representations (support drawings, notations and conversations), and can assist with remote collaborative design, are needed (Gabriel, 2000). At a basic level, a shared whiteboard can be an effective tool for remote designing (Nielsen, 1993). For example, Mailles-Viard Metz et al. (2015) developed a shared whiteboard called *SWHIFT* to support remote collaborative design. They conducted a collaborative design experiment to test its effectiveness also involving 42 architectural students. The results of their study suggest that, from several perspectives, collaboration in parametric design was able to overcome some of the problems faced by designers working in remote locations.

2.3.2. Augmented Reality

AR overlays an actual view of a real space or object with a computer-generated image or model in real-time (van Kleef et al., 2010). AR can also be used to describe the overlaying of artificial sounds on real ones, or simulate overlaps between any artificial and real sensory experience. In design, however, it is most commonly used to describe overlays of visual information. AR can be thought of as the opposite of telepresence. In telepresence, a user is projected into another place; in AR, additional information is artificially presented to the place of the user.

AR is typically realised using various types of screens. For example, Peddie (2017) considers three types: see through, obstructed and projected. In the see through category, a transparent device like smart glasses or a head-mounted device is used for displaying the AR information. At the same time the user can still see the real world through that device. In the obstructed AR view, a screen blocks the view of the real world content behind it to replace it with the augmented information (smart phones and tablets are the most common devices for this type of AR). In the third category, the digital content is projected into the real world, for example as holograms.

Another categorisation differentiates AR in terms of the breadth and orientation of the view. Most obstructed view devices cover only a limited view area (say a tablet in a person's hands). However, users can easily see the non-augmented reality around these small screens. This limited view may also apply to see through devices if they are not worn like eyeglasses or a head-mounted device. On the other hand, the device's screen may block the direct vision of reality and, therefore enable the overlay of the digital content in every viewing angle (this usually applies only to wearable devices and panoramic screens). Peddie's (2017) last category, the projected, is typically limited in size, although they can be three-dimensional and be viewed from various angles.

The third approach to categorising AR technology considers the distribution of the visualisation devices. In this regard, the first category comprises devices whose positioning and orientation are controlled by the creator of the AR content. These are usually stationary devices, like a holographic projector, large screens or panoramic walls. This type is suitable for situations where multiple people must simultaneously observe one form of information. In the second type, the digital content is visualised by separated devices held by individual users. While there may still be one server relaying the information, the final rendering is made by each device based on its individual orientation and position. In this case, the users have more control over what they might see. Table 2.1 summarises categories of various AR technologies.

Table 2.1. Categories of AR technologies.

Categories	Types	Examples of devices
Overlay	See through	Head-mounted devices, smart eyeglasses
	Obstructed	VR head-mounted devices, tablets, wall-screens, large TVs
	Projected	Holograms
View breadth	Full-angled (around person)	Head-mounted devices, eyeglasses, panoramic screens
	Full-angled (around content)	Holograms
	Limited	Tablets, wall screens
Rendering	User-based/mobile	Tablets, head-mounted devices
	Creator-based/ stationary	Holograms, panoramic rooms

AR systems require five components to properly generate their content. In addition to human stakeholders (creators and users) and the augmented real world itself, the five components are: non-present content, a tracking system, digital interpretation, a combiner and an output device (Manuri and Sanna, 2016).

The first component is the non-present content that augments the reality. This content is not necessarily artificial, it is just displaced from its usual location. The only required criteria is its absence from the natural view prior to the augmentation. Thus, AR could be used to overlay a real tree from one location in a park in a different location. In this example, both tree and park are real-world objects, but the AR is used to show how the tree will look or function in a different location.

The second component is a tracking system, being a set of sensory devices that detect the orientation and location of users or their devices. Manuri and Sanna (2016) identify two types of tracking systems: marker-based and marker-less. The former operates using pre-defined "markers" (such as QR codes) which provide information on the position of the viewer relative to the augmented subject. The marker-less version uses environmental information for this purpose. The two approaches can be further distinguished based on resources required. The common computationally low-cost approach processes geographic information like gyroscopic and geolocation data (Manuri and Sanna, 2016) to calculate the position of the viewer and the augmented item. In the high-cost approach, devices may contain technology to distinguish background scenes and objects using image recognition. Nevertheless, a tracking system is only necessary for mobile or user-based AR visualisation. In the stationary mode, the orientation and locations are pre-defined and fixed.

The third component is the digital interpretation of reality. This interpretation contains geometrical or geographical constraints that are necessary to properly overlay the artificial content on the real content. For example, if the purpose is to virtually reconstruct a historic building over its ruins, which are partially behind another building, the digital reconstruction should be in the background of the view. For this purpose, the easiest way is to include the geometrical data of the geographical context in the AR system (Fenais et al., 2019).

The fourth component is the "combiner", being the system that calculates, generates or positions the information necessary for the graphical compilation of the artificial content, and in case of obstructed overlays, some part of the actual reality. Specific software capabilities may differ between stationary and mobile rendering, this also needs to be considered. In mobile rendering, the software may need to update the rendering in real-time to match the user's movement and orientation. Additional features, like responsiveness to environmental variations (time, weather, or light for outdoors), may also be included in the software to enhance its realism.

Finally, there must be an output device that displays the superimposed content over the reality for the users. Typically, the output is generated in two stages, each of which may require a certain device or medium. The first is the generation of light particles representing the content. The second captures the particles on a medium where the human eye can see them. In devices with digital displays or obstructed overlays, both stages are performed on the same device. Tablets, wall-mounted displays and VR head-sets are examples of this type. In addition, there are also holographic "fans" with light emitters on their propellers. While the propeller spins too quickly for the human eye to detect, or its view to be hindered, the change in intensity and colour of the light can be read by the eye as a 2D image in the air. However, for other systems there must be a transparent medium (like lenses or particles in the air) to capture and reflect the light correctly. 3D glasses, optical head-mounted devices, smart eyeglasses and, digital contact lenses, possibly in the future are examples of these media. In some holographic projections, the medium is the air that reflects the light beams concentrated in certain locations based on the digital content.

Variations of AR applications

The application of AR has spread across many design industries and domains, from architecture to Mechanical, Electrical and Plumbing engineering (MEP), and from civil engineering to industrial design. Presently, many AR apps and platforms for portable devices are available for free at an affordable cost (Fraga-Lamas et al., 2018). These apps usually load a digital design model either directly from CAD or BIM tools, or as a saved file, and then based on some instruction by the user and their device's orientation, visualise the design of an image of the reality in the background. In this section, two general types of AR applications, including the content's visual type and chronology, are discussed and examples presented.

In general, an AR digital overlay is either contextually realistic or mostly concerned with data or information. The former involves a visual representation of an object as it would be located in a real place. The latter pertains to other forms of visual information which are not expected to have been in that place as a real object. An example of this distinction is the difference between the digitally reconstructed image of a historic building on its site, and text/images or other data appearing on screen to provide more information about the building and its past.

The next distinction between types of AR enhancements is chronological. The augmenting visuals can represent the past, present or future state of the augmented object. An example of the historic reconstruction is a past augmentative overlay. Present augmentations usually provide additional or currently invisible information (either real or unreal). For example, Pokemon Go, a popular game in the mid to late 2010s, superimposed images of fictional animals onto views of the real world around players. For future overlays, the AR system visualises the proposed or predicted states of the real-world objects or locations. For example, designers or producers may use AR visualisations to review design alternatives or introduce them to customers or clients. Finally, it is possible for augmentation to contain multiple stages, from past to present and possible future alternatives. For instance, in the context of repair and maintenance, the past, the augmented present, and the ideal or repaired future may all be visualised.

Finally, the scale and stage of the augmentation may vary in AR applications. In general, realistic digital content can be superimposed onto either the real life-size context or a real but schematic background. For the former, the presence of the AR user may be necessary in the actual location of the augmented place or object. However, for the latter, since the context is itself simulated, the user's presence in relation to the original context is irrelevant.

Design and AR

Each of the types of applications discussed previously can be used in AR in different industries, and for different purposes. Manuri and Sanna (2016) identified 10 areas where AR has been commonly applied, including medicine, assembly (and repair), entertainment (and sport), marketing, collaboration, cultural heritage, design, education and military. Five of these areas (design, assembly, collaboration, cultural heritage and education) are relevant in the context of the present book.

This section is focused on the application of AR in the areas of design marketing, cultural heritage and design process. Of these three, marketing-related communications are among the most commonly used applications of AR in the present day. These communications include the introduction of ideas or products to stakeholders (clients, customers, peers) for the sake of advertisement, trial, selection, or receiving feedback. A similar application involves collaboration between stakeholders during the design process (Fraga-Lamas et al., 2018; Manuri and Sanna, 2016). Considering the current high cost and limited availability of wearable AR devices, users commonly use a device like a tablet pointed at desired angles or markers to see the simulation of products or buildings in that place. Compared to more traditional methods – such as walk-throughs, static 2D rendering, 3D physical models and large printed sheets on construction sites – the AR overlay is generally more efficient in terms of cost and visualisation capacity. Nevertheless, despite these possibilities, the effectiveness of AR is debatable. For example, some surveys (Bulearca and Tamarjan, 2010; Connoly et al., 2010) indicate lower or comparable effectiveness compared to traditional print media, while others show only low-level effectiveness (Irshad and Rambli, 2011; Zulkifli et al., 2015). This difference may be related to technological advances or the limited scope of reviewed areas. A significant number of studies of AR advertisements focus on improving the quality of the visualisation and interaction without empirically examining the effectiveness of the advertisement itself (Rese et al., 2017). For example, among the nine cases surveyed by Fraga-Lamas et al. (2018), only two – an AR enhancement of catalogues (El-Firjani and Maatuk, 2016) and a different digital augmentation of shoes (Stoyanova et al., 2015) – performed in the analysis of user satisfaction. The first showed average-to-neutral results, and the second identified positive reactions, but neither compared the AR results with those of conventional product trials.

As mentioned previously, there are two scales of AR application: life-size and schematic. The schematic originals can be plans, catalogues, layouts or similar documents that illustrate an outline of the real design. The AR system then uses the outline to process the geometric and geographic information necessary for the visualisation. This approach has been applied in various fields, such as children's story books, entertainment or educational books, by augmenting the printed content with extra digital information (Cheng and Tsai, 2014; Düsner et al., 2012). In design, a commercial example has been developed by ELK, an Austrian housing design company, to allow its customers to view and customise prefabricated house options. The app uses a printed floor plan to position the 3D visualisation of the building. Another example is the experimental HBR AR app by PTC that creates a simulation of procedures and machinery based on printed schemes on the pages of an article in Harvard Business Review (HBR) journal published in 2017. Figure 2.3 shows an AR app called *Augment* being used to exhibit the housing model based on a floorplan.

An advantage of the scaled AR visualisation is that the user's perspective is typically an aerial overview of the presented visual. Not only is it possible to visualise a whole design in a small physical space from this perspective, it is also easier for a user to navigate around it. This is especially beneficial when there are multiple users who need to discuss and collaborate on a project. This approach also does not require all the users to be present in the same physical space, as long as they all have access to the same schematic context (Manuri and Sanna, 2016). An early example of this type of usage was *Studierstube* (Schmalstieg et al., 2002), a system through which multiple people could communicate with each other and interact with the augmented content.

The other category of scale-based AR applications is life-size, with a real background. This is usually achieved by users being present at the specific location. This type of AR is used less frequently for marketing or

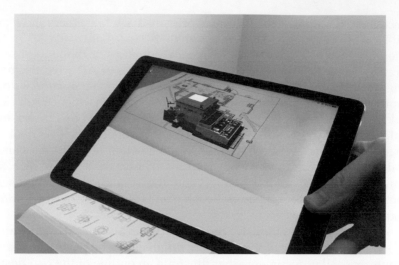

Figure 2.3. House previewing AR solution (Photo was taken by authors).

promotion purposes in engineering design, because the products of this industry are too context specific to be pre-designed, require physical usage or are too large to be properly visualised. For industrial design, however, selling furniture, appliances and decorative objects using in-situ previews is a valuable approach. Furniture visualisation using AR was considered in earlier research by Breen et al. (1996), and more recently, Khairnar et al. (2016) provided a survey of AR furniture preview applications and proposed several technological and interface enhancements. The most common approach used has been AR apps for mobile devices that place and preview furniture in people's homes (Wenbing, 2018). In addition to furniture, AR applications of this type have been used for advertising cars and automotive customisation, garments and fashion (Hauswiesner et al., 2013) and beauty products.

For architectural designers, the most common uses of AR are showcasing a building for a client or providing better representation of the design in its context. The apps used for these purposes are either bonded with mainstream CAD or BIM software, or use a standard CAD file format for visualisation. Autodesk *Revit* and *TrimbleSketchUp* are two CAD/BIM applications that have multiple AR plugins. For example, *Revit* can export AR with plugins, such as *Umbra3D* (*Umbra 3D*, 2019), WakeApp's *EnTiTi* (*EnTiTi*, 2017), *SightSpace Pro 3D* and *Kubity Pro* (Kubity, 2018). The last two also have plugins for *Sketchup*. Conversely, other CAD software use specific AR apps. For example, *AR-media* connects with Autodesk *Maya* and *3ds Max*. *Fologram* is specialised for Rhino 3D and Umbra3D also works with Graphisoft's *ArchiCAD*. Some AR apps like *SMACAR* (*SMACAR*, 2018), *AUGmentecture* (*Augmentecture* 2018) and *SightSpace* are also capable of uploading a standard 3D or CAD file independent of the

CAD software. In addition to these end-user apps, there are also various Software Development Kits (SDK) available for programmers to customise their own AR apps or presentations. Some of the apps mentioned in this section like *AR-media* and *Umbra3D* have their own specific SDKs. Other SDKs like *ARKit* and *Vuforia,* may be used with a rendering engine like *Unity 3D* for more realistic and efficient outputs.

Another common application of AR is for tourism and cultural heritage. Such examples normally function by overlaying digitally reconstructed content onto the real world. This application has been an area of interest for many disciplines, such as art history, archaeology and tourism. Before the emergence of AR, the most efficient methods for this purpose were 3D animations of historic reconstructions or scaled models. The use of AR for this purpose started in the late 1990s (Azuma et al., 2001; Feiner et al., 1997; van Gool et al., 1999). For example, a "touring machine" for the University of Columbia was developed by Feiner et al. (1997) as an early example of working AR. The system consisted of a backpack computer with a head-mounted display connected to it. In the early 2000s, the backpack AR system was developed to accommodate mobile AR interfaces such as *ARQuake* (Thomas et al., 2002) and *Archeoguide* (Vlahakis et al., 2002).

In the last two decades, with the increasing availability and affordability of tablets and smartphones, AR projections for historic contexts have become more common. Generally, the targets of the digital augmentation are either artefacts in museums, or historic reconstructions of buildings in their original locations. The former application is usually more accurate in representation, because in the indoor environment it is easier to associate the artefact with a marker that assists the AR system's orientation. For example, Stanco et al. (2012) and White et al. (2014) designed AR systems to visualise statues based on markers on paper. Gherardini et al. (2018) used the rectangular base of a damaged lion statue as the marker to find the viewer's position and visualise the missing parts of the statue. Nofal et al. (2018) used an image of the extant part of the historic artefacts in Nimrud, Iraq as the basis for orientating the reconstruction of the rest. In addition to these isolated examples, *GuidiGo* is a smartphone app that demonstrates the original state of artefacts or the way they were used or created in the past. *ARtGlass* (*ARtGlass*, 2019) has a similar function, although it uses wearable glasses. A common and relatively popular application of AR is in natural science museums, where AR helps to bring historic specimens to life by visualising their possible physical form over their displayed skeletons (Marques, 2012).

AR projections of historic ruins have also been demonstrated. One example, *ArcheoGuide* (Vlahakis et al., 2002), provides an overlay on the ruins of the Temple of Hera at Olympia in Greece. *ArcheoGuide* was similar to other early backpack devices with a head-mounted display, camera, a gyroscopic detector and a GPS receiver to pinpoint the location and orientation of the user. To increase the accuracy of the rendering, WiFi antennas were installed near the temple. Although the rendering was rudimentary, the main challenge with this system was ergonomics. *ArcheoGuide* is an example of the complexity of producing AR in the real world and in all weather conditions. Amongst the many difficulties with achieving this, Panou et al. (2018) emphasise the inaccuracy of GPS in consumer mobile devices as the main challenge. Even while using accelerometers, there is still a degree of inaccuracy in the overlay of the content. Accordingly, environmental factors such as lighting and weather may also undermine the usefulness of the visualisation. Another shortcoming is the lack of recognition of foreground objects and temporary intrusions (like people and cars) that stand between the historic background and the camera, creating an unrealistic view. Furthermore, due to the cost of creating detailed 3D models, the most common approach to the application of AR in exterior historical settings is photographic. In this approach the users are shown available photographs (or 2D illustrations) from certain angles. This can be achieved, for example, using *PIVOT* (*PIVOT*, 2019), an open-source app with an image database. A similar approach has been adopted by several cities; the Museum of London's app *StreetMuseum* and *VTT* for Helsinki (Rainio et al., n.d.) both superimpose historic photos over current views.

Application of AR for design tasks

The AR applications described in the previous section are all concerned with visualisation. Users typically can only control the designed alternative, process, or stage they see. However, some of the most advanced AR examples are able to demonstrate the operation or application of content in a dynamic way. This can be seen in the HBR machinery example or *GuidiGo*'s artefact usage demonstration. In the HBR machinery example, after creating a 3D virtual model of the machines on the tablet's screen, the software allows the user to schematically simulate the operational procedure of the machines by pushing various buttons or levers.

This interactive interface gives the user some capacity to manipulate content. In much the same way, some marketing applications for furniture offer visualisation options, like the size, texture or colour of the products. However, despite the existence of this interactive potential, it is rarely used in design. A possible explanation is that the engineering and industrial design disciplines are heavily reliant on numerical inputs and schematic representation. The former is easier with common devices, such as a keyboard, and the latter is antithetical to AR's common use for visualising realistic rather than schematic content. In this regard, design tasks with a more sculptural focus and less numerical dependencies could be more viable for AR applications. For example, Peng et al. (2018) developed

RoMA, a system that allows a user to design using AR, while a 3D printer creates a partial wireframe structure for the designer to use as a reference. Hattab and Taubin (2019) used AR to project carving instructions on a block of material so a sculptor has an accurate guide for cutting. Bosc et al. (2019) described a typical application of AR in reconstructive surgical processes. In this application, the intended form of the tissue or organ is projected using AR glasses to guide the surgeon. Valentini and Biancolini (2018) developed a haptic AR system for manipulating and sculpting 3D meshes that can be used as parts of products in engineering design. Finally, Jahn et al. (2019) developed an AR system that guided the bending of steel bars to create a 3D pavilion structure.

2.3.3. Virtual Reality

One of the key objectives of CAD development has been to remove any barrier between the designer and the design. This usually involves better or more intuitive control over inputs and a more realistic experience of outputs. Virtual Reality (VR) is potentially the ideal interface option for this purpose. VR generally refers to a digitally created, immersive environment (Milgram and Kishino, 1994; Sherman and Craig, 2018). In this sense, it differs from AR, as the AR user is still immersed in the real world.

Definitions of VR typically emphasise the need for the following seven components (Fuchs, 2017; Mihelj et al., 2013; Sherman and Craig, 2018). Firstly, there must be a virtual or artificial world that is computer-generated (this world may or may not resemble, feel and behave like the real world). Secondly, the creator of the virtual world sets the initial rules for its operations and may design some or all of the objects in it. Thirdly, participants in the virtual world can interact with it and its objects, potentially creating new objects or removing them. Fourthly, the VR interface and hardware allow the computer to detect human behaviours, like movement and orientation (a VR head-mounted device, for example, records the direction of vision and the general location of the viewer, and VR gloves record the position or movement of the user's arms, hands and fingers). Fifthly, sensory feedback simulates the responses that a person inside it would have received if that virtual world was real (the most common feedback methods for this purpose are visual and aural, simulating the sights and sounds of the artificial world, however, limited haptic feedback is also possible, as is kinematic feedback for orientation and gravity in some flight simulators). Next, real-time interaction is an essential component of VR, which requires an almost instantaneous response time. For example, if the persons are walking in VR, the virtual world projection around them must be continuously updated based on their position and orientation. If the person is able to manipulate or modify certain objects in the virtual world, the changes would take place and be presented in real-time. Finally, the seventh component is virtual "presences"; this is the sensation that a person is present and engaged in the virtual world (the sense of immersion or presence is a subjective matter, nevertheless, it is possible to identify factors influencing it, the quality of the interaction and response time are important factors).

In addition to these seven components, Sherman and Craig (2018) argue that an effective virtual world has a level of vividness that evokes a deep sense of an alternative reality. Higher levels of sensory feedback, which cover a broad range of senses, support the creation of vivid virtual worlds. Sherman and Craig (2018) also emphasise that the capacity to interact with the world has an impact on its vividness. They identify three properties of effective interaction. First, the interaction needs to occur in real-time. Second, the capacity to modify the environment must provide a breadth of choices. Finally, the mapping of interactions in the virtual world must reflect some of the diversity present in the real world. Ultimately, the sense of presence (the final component listed previously) depends on the vividness of the virtual world and the predictability and sensibility of these interactive relationships it accommodates.

VR paradigms, devices, media and applications

VR is concerned with immersive representation. Historically, cave paintings and murals lined the walls of early shelters, evoking the experience of a space that is not physically present. Novels and poetry also provided readers with different types of immersive experiences, as too did the first Gothic and baroque cathedrals, which were arguably the most immersive virtual realities of their times. In the late 19th century, the very first audiences of projected films were genuinely scared or shocked by seeing the projection of a train coming towards them. However, none of these experiences featured the possibility of the audience interacting with the content, which is a key component of VR. Another component of effective VR, the non-present world's realistic impact on the audience, was simulated using various types of technology. In the 1930s, Aldous Huxley imagined devices that could transfer feelings, as well as picture and sound, in movies, or as he called them, "feelies". An early attempt to realise this idea was the 1957 *Sensorama* project that simulated riding a bike through the streets of New York.

One of the first serious attempts to create an immersive and interactive virtual environment was *The Sword of Damocles*. This system comprised a head-mounted visual display and movement orientation device, suspended from a ceiling-mounted frame above the user's head, like the titular sword of Greek mythology. It was a precedent for the head-based paradigm of VR technologies that has since become the international standard (Sherman and Craig, 2018). This paradigm features a head-mounted device that projects virtual imagery directly into the wearer's eyes. In addition, the head-mounted device tracks the movement and orientation of the head to update the visual feedback accordingly. Early head-mounted devices were mechanically connected to a set of arms that processed the data of head movements. Modern devices, however, use a variety of miniaturised integrated sensors to detect motion, rotation and direction, with wireless communication to the main system.

One negative side effect of the use of a head-mounted device for VR involves loss of visual connection to the body and sometimes subsequent imbalance or nausea. To mitigate that, some modern VR includes representations of the user's body, especially hands or legs, in the simulated world. Hands, in particular, have been frequent subjects of representational attempts, not only to improve the sense of presence, but also to emphasise the capacity for interaction. For this purpose, the most common devices have been gesture-detecting "gloves" and multi-directional handles or levers. Like the head-mounted device, the older gloves consisted of relatively large mechanical arms that detected the position, direction and angle of finger joints in addition to the movements of the forearms. The handle or lever devices were typically used for simulating tools and machines in the virtual world. They may have used buttons or defined gestures to trigger different functions. Another type of device used for this purpose is the handheld controller, which is commonly used in gaming and entertainment.

The second paradigm in VR devices was inspired by cinematic approaches. In this paradigm, instead of the head-mounted device being the projector, the virtual content is projected or displayed on a large display or surfaces of the room in which the users are located. The main difference in the experience of presence between this paradigm and the former is that the users in the latter retain a visual connection with their own body. However, in this paradigm it is not possible to realistically immerse their hands or tools in the virtual world, and thus manipulation of objects is less realistic, or is limited to devices that have less localised effects, like simulated pointers or guns. Another difference is that the former paradigm, the head-mounted device, can provide the opportunity for users to physically move in the real world. However, the latter paradigm only allows limited movements within the field of projection. In addition, as the virtual world is already rendered on the room surfaces in a panoramic way, head orientation does not affect the virtual camera view. In contrast, rendering in the head-mounted device must be updated constantly to respond to orientation and head angle. For these reasons, the former paradigm is sometimes called *mobile* and the latter *stationary*. Nevertheless, in both paradigms the options for movement are limited. Devices like stationary walking surfaces (similar to treadmills), controlling mechanical devices such as pedals or steering wheels, and handheld controllers can be used in combination to simulate realistic movements.

The third paradigm in VR is fundamentally different from the previous two, because it does not simulate a panoramic view around the person. Instead, the virtual content is rendered on a smaller screen like that of a tablet. Many such devices possess movement, orientation and location sensors allowing them to create simple VR environments.

The earliest attempts to create a VR experience were developed for the military and entertainment sectors. Today, the games industry has driven the development of affordable head-mounted technologies. VR games range from war games to sports, racing, flight simulation and even design. Designing and drafting were the third main area where VR has been applied, although to date the majority of developments have been focused on visualising the designed product, not the act of designing itself. A motivating factor for using VR for visualisation is to better examine the function and appearance of the design prior to its costly production (Fuchs, 2017; Mihelj et al., 2013). Another justification is that the VR environment potentially provides a more thorough experience than older presentation methods such as scaled models, sketches and 2D rendering. In addition to military, entertainment and design applications, VR has also been used for educational purposes. In particular, VR offers an ideal platform for training activities that may otherwise be costly or may cause risk to health or property (Mihelj et al., 2013), and empowering virtual classrooms where people can take part in a shared educational experience, regardless of where they are in the world. Finally, VR can be used to create encouraging environments for rehabilitating patients with specific motor disabilities or psychological needs, without exposing them to risks or requiring extensive real-world resources.

2.4. Conclusion

This chapter has reviewed the main emergent technologies in computational design: CAD, parametric and generative design, telepresence, augmented reality and virtual reality. The review has focused on key concepts, historic developments, applications and recent research. From its origins in mid-1950s CAD, computational design has evolved through 2D drafting to 3D modelling, rule algorithm scripting, virtual and augmented reality. Some of the technologies such as CAD, have been applied in AEC industries and various design-related fields for several decades, thus they are relatively mature and widely applied. Others, such as parametric and generative design, are the focus of ongoing research, and have only been applied in industry in recent years. Those technologies still have extensive potential for further study.

At the start of this chapter we noted that historically, for example, pens replaced graphite and transparent tracing paper replaced papyri, but the general use of these tools and media remained largely unchanged over time. However as this chapter reveals, emergent computational design technologies can potentially change the way designers work and even think. That is the topic of the next chapter, which looks at design thinking and computational design thinking, and begins to ask "will the computer assist or hinder creativity in design?"

References

Abdelsalam, M. (2009). The use of the smart geometry through various design processes: Using the programming platform (parametric features) and generative components. *In: Proceedings of the Arab Society for Computer Aided Architectural Design (ASCAAD 2009)*, Manama, Kingdom of Bahrain. http://papers.cumincad.org/cgi-bin/works/paper/ascaad2009_mai_abdelsalam

Agkathidis, A. (2015). Generative design methods—Implementing computational techniques in undergraduate architectural education. *In: Proceedings of the 33rd eCAADe Conference* (pp. 297–304). Vienna, Austria.

Almusharaf, A.M. and Elnimeiri, M. (2010). A performance-based design approach for early tall building form development. *In: Proceedings of the Arab Society for Computer Aided Architectural Design (ASCAAD 2010)*, Fez, Morocco.

Arora, J., Jain, V., Saini, A., Shrey, S., Mehra, N. and Parnami, A. (2019). VirtualBricks: Exploring a scalable, modular toolkit for enabling physical manipulation in VR. pp. 1-12. *In: CHI '19: Proceedings of the 2019 CHI Conference on Human Factors in Computing Systems*, Glasgow, Scotland (Paper No. 56). https://doi.org/10.1145/3290605.3300286

Augmentecture (2018). Retrieved 08/07/2019 from https://www.augmentecture.com

Azuma, R., Baillot, Y., Behringer, R., Feiner, S., Julier, S. and MacIntyre, B. (2001). Recent advances in augmented reality. *IEEE Computer Graphics and Applications*, 21(6), 34–47.

Baerlecken, D., Martin, M., Judith, R. and Arne, K. (2010). Integrative parametric form-finding processes. *In: Proceedings of the 15th International Conference on Computer Aided Architectural Design Research in Asia (CAADRIA)*, Hong Kong.

Batty, M. (1997a). Cellular automata and urban form: A primer. *Journal of American Planning Association*, 63(2), 266–274.

Batty, M. (1997b). Urban systems as cellular automata. *Environment and Planning B: Planning and Design*, 24, 159–164.

Bernal, M. (2011). Analysis model for incremental precision along design stages. *In: Proceedings of the 16th International Conference on Computer Aided Architectural Design Research in Asia (CAADRIA)*, Newcastle, Australia.

Blum, C. and Li, X. (2008). Swarm intelligence in optimization. pp. 43–85. *In:* C. Blum and D. Merkle (Eds.), Swarm Intelligence. Springer.

Boden, M. and Edmonds, E. (2007). What is generative art? *Digital Creativity*, 20(1), 21–46.

Bosc, R., Fitoussi, A., Hersant, B., Dao, T.H. and Meningaud, J.P. (2019). Intraoperative augmented reality with heads-up displays in maxillofacial surgery: A systematic review of the literature and a classification of relevant technologies. *Journal of Oral and Maxillofacial Surgery*, 48, 132–139.

Breen, D.E., Whitaker, R.T., Rose, E. and Tuceryan, M. (1996). Interactive occlusion and automatic object placement for augmented reality. *Computer Graphics Forum*, 15(3), 11–22.

Briggs, S.J. and Srinivasan, M.A. (2002). Haptic interfaces. pp. 93–115. *In*: K.S. Hale and K.M. Stanney (Eds.), Handbook of Virtual Environments: Design, Implementation, and Applications. CRC Press.

Bulearca, M. and Tamarjan, D. (2010). Augmented reality: A sustainable marketing tool? *Global Business and Managment Research*, 2(3), 237–252.

Burry, J. and Holzer, D. (2009). Sharing design space: Remote concurrent shared parametric modeling. *Proceedings of 27th eCAADe Conference*, Istanbul, Turkey.

Burry, M. (2003). Between intuition and process: Parametric design and rapid prototyping. pp. 149–162. *In*: B. Kolarevic (Ed.), Architecture in the Digital Age—Design and Manufacturing. Spon Press.

Cárdenas, C.A. (2007). Modeling Strategies: Parametric Design for Fabrication in Architectural Practice [Doctoral dissertation]. Harvard.

Chakrabarti, A., Shea, K., Stone, R., Cagan, J., Campbell, M., Hernandez, N.V. and Wood, K.L. (2011). Computer-based design synthesis research: An overview. *Journal of Computing and Information Science in Engineering*, 11, 021003-021012.

Chamaret, D., Ullah, S., Richard, P. and Naud, M. (2010). Integration and evaluation of haptic feedbacks: From CAD models to virtual prototyping. *International Journal of Design Manufacturing*, 4, 87–94.

Cheng, K.-H. and Tsai, C.-C. (2014). Children and parents' reading of an augmented reality picture book: Analyses of behavioral patterns and cognitive attainment. *Computers & Education*, 72(2014), 302–312.

Connoly, P., Chambers, C., Eagleson, E., Matthews, D. and Rofers, T. (2010). Augmented reality effectiveness in advertising. *In*: *Proceedings of the 65th Midyear Conference on Engineering Design Graphics Division of ASEE*. ASEE.

Dal Co, F. (2003). The world turned upside down: The tortoise flies and the hare threatens the lion. pp. 39–61. *In*: D.C. Francesco and W.F. Kurt (Eds.), Frank O. Gehry: The Complete Works. New York: Monacelli Press.

Davis, M.R. and Ellis, T.O. (1964). *The Rand Tablet: A Man-machine Graphical Communication Device* (189–61). https://www.rand.org/pubs/research_memoranda/RM4122.html

Donald, T. (1962). A Sumerian plan in John Rylands Library. *Journal of Semitic Studies*, 7(2), 184–190.

Düsner, A., Walker, L., Horner, H. and Bentall, D. (2012). Creating interactive physics education books with augmented reality. *Proceedings of the 24th Australian Computer–Human Interaction Conference*. Melbourne, pp. 107–114. https://doi.org/10.1145/2414536.2414554

Eastman, C.M. (2008). *BIM Handbook: A Guide to Building Information Modeling for Owners, Managers, Designers, Engineers and Contractors*. Wiley. http://books.google.com.au/books?id=IioygN0nYzMC

Eckert, C., Kelly, I. and Stacey, M. (1999). Interactive generative systems for conceptual design: An empirical perspective. *Artificial Intelligence for Engineering Design, Analysis and Manufacturing*, 13, 303–329.

El-Firjani, N.F.M. and Maatuk, A.M. (2016). Mobile augmented reality for interactive catalogue [Conference paper]. 2016 International Conference on Engineering & MIS (ICEMIS), Agadir, Morocco.

Englebert, D.C. (1970). *X-Y Position Indicator for a Display System* (United States Patent No. US003541541). U. S. P. a. T. Office.

EnTiTi AR/VR Exporter Revit. (2017). Retrieved 08/07/2019 from https://apps. autodesk.com/RVT/en/Detail/Index?id=2453573418981504717&appLang=en&os=Win64

Fasoulaki, E. (2007). *Genetic Algorithms in Architecture: A Necessity or a Trend?* Generative Art 2007, Milan, Italy.

Feiner, S., MacIntyre, B., Höllerer, T. and Webster, A. (1997). A touring machine: Prototyping 3D mobile augmented reality systems for exploring the urban environment. *International Symposium on Wearable Computing*, pp. 74–81. IEEE.

Fenais, A., Ariaratnam, S.T., Ayer, S.K. and Smilovsky, N. (2019). Integrating geographic information systems and augmented reality for mapping underground utilities. *Infrastructures*, 4(60).

Fischer, M., Burry, M. and Frazer, J. (2003). Triangulation of generative form for parametric design and rapid prototyping. *Proceedings of 21th eCAADe Conference*, Graz, Austria.

Fischer, T. and Herr, C. (2000). Teaching generative design. *In*: C. Soddu (Ed.). *Proceedings of the 4th International Conference on Generative Art*. http://papers. cumincad.org/cgi-bin/works/Show?c78f

Forster, K.W. (1998). *Frank O. Gehry: Guggenheim Museum Bilbao*. London: Edition Axel Menges.

Frazer, J. (2016). Parametric computation: History and future. *Architectural Design*, 86(2), 18–23.

Fuchs, P. (2017). *Virtual reality headsets—A theoretical and pragmatic approach*. CRC Press, Taylor and Francis.

Gabriel, G. (2000). Computer Mediated Collaborative Design in Architecture: The Effects of Communication Channels in Collaborative Design Communication [Doctoral dissertation]. University of Sydney.

Gane, V. and Haymaker, J. (2009). Design scenarios: Methodology for requirements driven parametric modeling of high-rises. *In*: X. Wang and N. Gu (Eds.), *Proceedings of the 9th International Conference on Construction Applications of Virtual Reality* (CONVR 2009), Sydney, Australia.

Garnier, S., Gautrais, J. and Theraulaz, G. (2007). The biological principles of swarm intelligence. *Swarm Intelligence*, 1, 3–31.

Gero, J.S. (1996). Creativity, emergence and evolution in design. *Knowledge-Based Systems*, 9(7), 435–448. https://doi.org/10.1016/s0950-7051(96)01054-4

Gero, J.S. and Kazakov, V. (1996). An exploration-based evolutionary model of generative design process. *Computer Aided Civil and Infrastructure Engineering*, 11(3), 211–218.

Gherardini, F., Santachiara, M. and Leali, F. (2018). 3D virtual reconstruction and augmented reality visualization of damaged stone sculptures. IOP Conference Series: *Materials Science and Engineering*, Vol. 364. https://doi. org/10.1088/1757-899X/364/1/012018

Grasl, T. and Economou, A. (2013). From topologies to shapes: Parametric shape grammars implemented by graphs. *Environment and Planning B: Planning and Design*, 40, 905–922.

Gu, N. and Ostwald, M.J. (2012). Computational methods and technologies: Reflections on their impact on design and education. *In*: N. Gu and X. Wang (Eds.), Computational Design Methods and Technologies: Applications in CAD, CAM and CAE Education. IGI Global.

Hall, D.L. (1986). Computer-aided Drafting for Scenic and Lighting Designers: A Training Guide Using the Autocad® Drafting Package and the Zenith Z-100 Microcomputer [Doctoral dissertation]. University of Florida.

Hattab, A. and Taubin, G. (2019). Rough Carving of 3D Models with Spatial Augmented Reality [Conference paper]. ACM Symposium on Computational Fabrication 2019.

Hauswiesner, S., Straka, M. and Reitmayr, G. (2013). Virtual try-on through image-based rendering. *IEEE Transactions on Visualization and Computer Graphics*, 19(9), 1552–1565.

Hernandez, C.R.B. (2006). *Design Procedure: A Computational Framework for Parametric Design and Complex Shapes in Architecture*. Massachusetts Institute of Technology.

Herr, C. and Kvan, T. (2007). Adapting cellular automata to support the architectural design process. *Automation in Construction*, 16(2007), 61–69.

Hillyard, R.C. and Braid, I.C. (1978). Analysis of dimensions and tolerances in computer-aided mechanical design. *Computer-Aided Design*, 10(3), 161–166. https://doi.org/10.1016/0010-4485(78)90140-9

Hnizda, M. (2009). Systems-thinking: Formalization of parametric process. *Proceedings of the Arab Society for Computer Aided Architectural Design (ASCAAD 2009), 4th International Conference*, pp. 215–223. http://papers.cumincad.org/cgi-bin/works/Show?ascaad2009_marek_hnizda

Holland, J. (1975). *Adaptation in Natural and Artificial Systems: An Introductory Analysis with Application to Biology, Control, and Artificial Intelligence*. University of Michigan Press.

Holzer, D., Hough, R. and Burry, M. (2007). Parametric design and structural optimisation for early design exploration. *International Journal of Architectural Computing*, 5(4), 625–643.

Hornby, G.S. and Pollack, J. (2001). The advantages of generative grammatical encodings for physical design. *In*: *Proceedings of the 2001 IE Congress on Evolutionary Computation*, pp. 600–607. IEEE.

Irshad, S. and Rambli, D.R.A. (2011). User experience satisfaction of mobile-based AR advertising applications. *In*: H. Badioze Zaman et al. (Eds.), Advances in Visual Informatics. 4th International Visual Informatics Conference Proceedings, IVIC 2015. Lecture Notes in Computer Science, Vol. 9429, pp. 432–442. Springer.https://doi.org/10.1007/978-3-319-25939-0_38

Jahn, G., Newnham, C., van den Berg, N. and Beanland, M. (2019). Making in mixed reality. pp. 88–97. *In*: P. Anzalone, M. del Signore and A.J. Wit (Eds.), Recalibration: On Imprecision and Infidelity. *Proceedings of the 38ᵗʰ Annual Conference of the Association for Computer Aided Design in Architecture (ACADIA 2018)* ACADIA.

Jo, J. and Gero, J.S. (1998). Space layout planning using an evolutionary approach. *Artificial Intelligence for Engineering*, 12(3), 149–162.

Johnson, T.E. (1963). Sketchpad III: A computer program for drawing in three dimensions. *In: AFIPS '63 (Spring): Proceedings of the May 21–23, 1963, Spring Joint Computer Conference*. Association for Computing Machinery. https://doi.org/10.1145/1461551.1461592

Karle, D. and Kelly, B. (2011). Parametric thinking. *In*: J. Cheon, S.N. Hardy and T. Hemsath (Eds.), *Proceedings of ACADIA Regional 2011 Conference*, pp. 109–113.

Khairnar, K., Khairnar, K., Mane, S. and Chaudhari, R. (2016). Furniture layout application based on marker detection and using augmented reality. *International Journal of Engineering Science and Computing*, 6(4), 4201–4204.

Knight, T. (2003). Computing with emergence. *Environment and Planning B, Planning and Design*, 30, 125–156.

Kolarevic, B. (2003). *Architecture in the Digital Age: Design and Manufacturing*. Spon Press.

Krish, S. (2011). A practical generative design method. *Computer-Aided Design*, 43, 88–100.

Kubity | The Photoreal AR/VR Machine. (2018). https://pro.kubity.com

Lai, Y. (2017). *Augmented Reality Visualization of Building Information Model* [Degree Master of Science]. The Ohio State University.

Larsson, A. (2007). Banking on social capital: Towards social connectedness in distributed engineering design teams. *Design Studies*, 28(6), 605–622. https://doi.org/10.1016/j.destud.2007.06.001

Lazzeroni, C., Bohnacker, H., Gross, B. and Laub, J. (2012). *Generative Design: Visualize, Program, and Create with Processing*. Princeton Architectural Press.

Lee, J.H., Ostwald, M. and Gu, N. (2016). A Justified Plan Graph (JPG) grammar approach to identifying spatial design patterns in an architectural style. *Environment and Planning B: Urban Analytics and City Science*, 45(1), 67–89. https://doi.org/10.1177/0265813516665618

Light, R. and Gossard, D. (1982). Modification of geometric models through variational geometry. *Computer Aided Design*, 14(4), 209–214.

Maher, A. and Burry, M. (2003). The parametric bridge: Connecting digital design techniques in architecture and engineering. *Proceedings of the 2003 Annual Conference of the Association for Computer Aided Design in Architecture*, pp. 39–47. http://papers.cumincad.org/cgi-bin/works/Show?acadia03_005

Maher, M.L. (1990). Process models for design synthesis. *AI Magazine*, 11, 49–58.

Mailles-Viard Metz, S., Marin, P. and Vayre, E. (2015). The shared online whiteboard: An assistance tool to synchronous collaborative design. *European Review of Applied Psychology*, 65(5), 253–265. https://doi.org/https://doi.org/10.1016/j.erap.2015.08.001

Manuri, F. and Sanna, A. (2016). A survey on applications of augmented reality. *Advances in Computer Science: An International Journal*, 5(1), 18-27.

Marques, D.C.V. (2012). The Visitor Experience Using Augmented Reality on Mobile Devices in Museum Exhibitions [Doctoral dissertation]. Universidade do Porto. Porto, Portugal.

Mihelj, M., Noval, D. and Beguš, S. (2013). *Virtual Reality Technology and Applications*. Springer.

Milgram, P. and Kishino, F. (1994). A taxonomy of mixed reality visual displays. *EICE Transactions on Information and Systems*, E77-D (12), 1321–1329.

Monedero, J. (2000). Parametric design: A review and some experiences. *Automation in Construction*, 9(4), 369–377.

Muñoz-La Rivera, F., Vielma, J.C., Herrera, R.F. and Carvallo, J. (2019). Methodology for building information modeling (BIM) implementation in structural engineering companies (SECs). *Advances in Civil Engineering.* https://doi.org/- 10.1155/2019/8452461

Nielsen, J. (1993). *Usability Engineering.* Academic Press.

Nofal, E., Elhanafi, A., Hameeuw, H. and Vande Moere, A. (2018). Architectural contextualization of heritage museum artifacts using augmented reality. *Studies in Digital Heritage*, 2(1), 42–67.

Ostwald, M. (2004). Freedom of form: Ethics and aesthetics in digital architecture. *The Philosophical Forum: Special issue on Ethics and Architecture*, XXXV(2), 201–220.

Ostwald, M.J. (2006). *The Architecture of the New Baroque: A Comparative Study of the Historic and New Baroque Movements in Architecture.* Singapore: Global Arts.

Ostwald, M.J. (2010). Ethics and the auto-generative design process. *Building Research and Information: International Research, Development, Demonstration and Innovation*, 38(4), 390–400.

Ostwald, M. (2012). Systems and enablers: Modeling the impact of contemporary computational methods and technologies on the design process. pp. 1–17. *In*: N. Gu and X. Wang (Eds.), Computational Design Methods and Technologies: Applications in CAD, CAM and CAE Education. IGI Global.

Ostwald, M.J. (2015). Ethics and geometry: Computational transformations and the curved surface in architecture. *In*: K. Williams and M.J. Ostwald (Eds.), Architecture and Mathematics: From Antiquity to the Future. Volume II: 1500s to the Future. Birkhäuser/Springer.

Ostwald, M.J. (2017). Digital research in architecture: Reflecting on the past, analysing the trends, and considering the future. *Architectural Research Quarterly*, 21(4), 351–358. https://doi.org/10.1017/S135913551800009X

Oxman, R. (1990). Design shells: A formalism for prototype refinement in knowledge-based design systems. *Artificial Intelligence for Engineering*, 5(1), 2–8.

Panou, C., Ragia, L., Dimelli, D. and Mania, K. (2018). An architecture for mobile outdoors augmented reality for cultural heritage. *International Journal of Geo-Information*, 2018(7), 463–486.

Parish, Y. and Müller, P. (2001). Procedural modeling of cities. pp. 301–308. *In*: *Proceedings of the 28th Annual Conference on Computer Graphics and Interactive Techniques.*

Peddie, J. (2013). *History of Visual Magic in Computers: How Beautiful Images are Made in CAD, 3D, VR and AR.* Springer.

Peddie, J. (2017). *Augmented Reality: Where We Will All Live.* Springer.

Peng, H., Briggs, J., Wang, C.-Y., Guo, K., Kider, J., Mueller, S., Baudisch, P. and Guimbretière, F. (2018). RoMA: Interactive fabrication with augmented reality and a robotic 3D printer. pp. 1–12. *In*: *Proceedings of the 2018 CHI Conference on Human Factors in Computing Systems* (Paper No. 579). https://doi.org/10.1145/3173574.3174153

PIVOTTheWorld: Engage history, change your point of view. (2019). Retrieved 20/08/2019 from https://www.pivottheworld.com

Prince, M.D. (1966). Man-computer graphics for computer-aided design. *Proceedings of the IEEE*, 54(12), 1698–1708. https://doi.org/10.1109/PROC.1966.5251

Rainio, K., Honkamaa, P. and Spilling, K. (n.d.). *Presenting Historical Photos Using Augmented Reality.* http://virtual.vtt.fi/virtual/proj2/multimedia/media/publications/Presenting_Historical_Photos_in_Augmented_Reality.pdf

Rajus, V.S., Woodbury, R., Erhan, H., Riecke, B.E. and Mueller, V. (2010). Collaboration in parametric design: Analyzing user interaction during information sharing. *Life information. Proceedings of the 30th Annual Conference of the Association for Computer Aided Design in Architecture (ACADIA)*, pp. 320–326.

Rese, A., Baier, D., Geyer-Schulz, A. and Schreiber, S. (2017). How augmented reality apps are accepted by consumers: A comparative analysis using scales and opinions. *Technological Forecasting and Social Change*, 124(2017), 306.

Ross, D.T. (1960). *Computer-aided Design: A Statement of Objectives* [Technical Memorandum]. MIT.

Ruffle, S. (1986). Architectural design exposed: From computer-aided drawing to computer-aided design. *Environment and Planning B: Planning and Design*, 13(4), 385–389. https://doi.org/10.1068/b130385

Schlueter, A. and Thesseling, F. (2008). Balancing design and performance in building retrofitting: A case study based on parametric modeling. *Silicon + Skin: Biological Processes and Computation. Proceedings of the 28th Annual Conference of the Association for Computer Aided Design in Architecture (ACADIA)*, pp. 214–221.

Schmalstieg, D., Fuhrmann, A., Hesina, G. and Szalavári, Z. (2002). The Studierstube Augmented Reality Project. *PRESENCE: Virtual and Augmented Reality*, 11(1), 33–54.

Schnabel, M. (2008). *Mixed Reality in Architecture, Design and Construction.* Springer.

Schnabel, M.A. (2007). Parametric designing in architecture. pp. 237–250. In: A. Dong, A. Vande Moere and J.S. Gero (Eds.), *CAADFutures'07: Proceedings of the 12th International Conference on Computer Aided Architectural Design Futures.* Springer.

Schumacher, P. (2009). Parametricism —A new global style for architecture and urban design. *AD Architectural Design—Digital Cities*, 79(4), 14–23.

Seifi, H., Fazlollahi, F., Oppermann, M., Sastrillo, J.A., Ip, J., Agrawal, A., Park, G., Kuchenbecker, K.J. and MacLean, K.E. (2019). Haptipedia: Accelerating Haptic Device Discovery to Support Interaction and Engineering Design [Conference paper]. *Proceedings of the 2019 CHI Conference on Human Factors in Computing Systems.* Glasgow, Scotland.

Shaer, O. and Hornecker, E. (2009). Tangible user interfaces: Past, present, and future directions. *Human–Computer Interaction*, 3(1–2), 1–137.

Sherman, W.R. and Craig, A.B. (2018). *Understanding Virtual Reality: Interface, Application, and Design.* Morgan Kaufmann.

Singh, V. and Gu, N. (2011). Towards an integrated generative design framework. *Design Studies*, 33, 185–207.

Sarkar, P. (2000). A brief history of cellular automata. *ACM Comput. Surv.*, 32, 80–107. https://doi.org/10.1145/349194.349202

SMACAR—Augmented Reality Android App. (2018). Retrieved 08/07/2019 from https://play.google.com/store/apps/details?id=com.yakshamobileapps.smacar

Stanco, F., Tanasi, D., Gallo, G., Buffa, M. and Basile, B. (2012). Augmented perception of the past—The case of Hellenistic Syracuse. *Journal of Multimedia*, 7(2), 211–216.

Stiny, G. (1976). Two exercises in formal composition. *Environment and Planning B: Planning and Design*, 3(2), 187–210. https://doi.org/10.1068/b030187

Stiny, G. and Gips, J. (1972). Shape Grammars and the Generative Specification of Painting and Sculpture. *Proceedings of Information Processing 71.* Amsterdam, pp. 1460–1465.

Stiny, G. and Mitchell, W.J. (1978). The Palladian grammar. *Environment and Planning B*, 5, 5–18.

Stoyanova, J., Brito, P.Q., Georgieva, P. and Milanova, M. (2015). Comparison of Consumer Purchase Intention between Interactive and Augmented Reality Shopping Platforms through Statistical Analyses [Conference paper]. *2015 International Symposium on Innovations in Intelligent SysTems and Applications* (INISTA), Madrid, Spain.

Sundaram, B.L. (2017). The comparative study on the technology advancement and inventions in the Indus Valley and the Egyptian civilisation. *International Journal of Research and Analytical Reviews*, 4(3), 92–96.

Sutherland, I. (1963). *Sketchpad: A Man-Machine Graphical Communication System.* Massachusett's Institute of Technology.

Tasheva, S.B. (2012). Semiotic Aspects of Architectural Graphics' History [Conference paper]. The 10th World Congress of the International Association for Semiotic Studies, La Coruna, Spain.

Tching, J., Reis, J. and Paio, A. (2016). A cognitive walkthrough towards an interface model for shape grammar implementations. *Computer Science and Information Technology*, 4, 92–119.

Thomas, B., Close, B., Donoghue, J., Squires, J., De Bondi, P. and Piekarski, W. (2002). First person indoor/outdoor augmented reality application: ARQuake. *Personal and Ubiquitous Computing*, 6(2002), 75–86.

Umbra—Next-generation 3D tech for the web (2019). Retrieved 08/07/2019 from https://www.umbra3d.com

Valentini, P.P. and Biancolini, M.E. (2018). Interactive sculpting using augmented reality, mesh morphing, and force feedback: Force-feedback capabilities in an augmented reality environment. *IEEE Consumer Elecronic Magazine*, 7(2), 83–90.

Van Gool, L., Tuytelaars, T. and Pollefeys, M. (1999). Adventurous Tourism for Couch Potatoes. pp. 98–107. *In*: F. Solina and A. Leonardis (Eds.), Computer Analysis of Images and Patterns: CAIP 1999. Lecture Notes in Computer Science, Vol 1689. Springer. https://doi.org/10.1007/3-540-48375-6_13

Van Kleef, N., Noltes, J. and van der Spoel, S. (2010). *Success Factors for Augmented Reality Business Models.* https://www.inter-actief.utwente.nl/studiereis/pixel/files/indepth/KleefSpoelNoltes.pdf

Venkatraman, N. (1994). IT-enabled business transformation: From automation to business scope redefinition. *Sloan Management Review*, 35(2), 73–87.

Vlahakis, V., Ioannidis, N., Karigiannis, J., Tsotros, M., Gounaris, M., Stricker, D., Gleue, T., Daehne, P. and Almeida, L. (2002). Archeoguide: An augmented reality guide for archaeological sites. *IEEE Computer Graphics and Applications*, 22(5), 52–60.

von Neumann, J. (1951). The general and logical theory of automata. *In*: A.H. Taub (Ed.), John von Neumann: Collected Works. Vol. V, pp. 280–326. Pergammon Press.

Weisberg, D.E. (2008). *The Engineering Design Revolution.* https://www.cadhistory. net/toc.htm

Wenbing, H.S. (2018). *Online Furniture Shopping Using Augmented Reality.* Universiti Tunku Abdul Rahman. Kampar, Malaysia.

White, M., Liarokapis, F., Darcy, J., Mourkoussis, N., Petridis, P. and Lister, P.F. (2014). Augmented reality for museum artefact visualization. pp. 75–80. *In: Proceedings of the Fourth Irish Workshop on Computer Graphics.* Eurographics Ireland Chapter, Coleraine, Ireland.

Wolfram, S. (1986). Random sequence generation by cellular automata. *Advances in Applied Mathematics,* 7, 123–169.

Woodbury, R., Aish, R. and Kilian, A. (2007). Some patterns for parametric modeling. pp. 222–229. *In: Proceedings of the 27th Annual Conference of the Association for Computer Aided Design in Architecture.* Halifax, Nova Scotia.

Woodbury, R. (2010). *Elements of Parametric Design.* Routledge. http://books. google.com.au/books?id=HIM3QAAACAAJ

Yang, D., Sun, Y., Stefano, D.D. and Turrin, M. (2017). *A Computational Design Exploration Platform Supporting the Formulation of Design Concepts.* Society for Computer Simulation International.

Zellner, P. (1999). *Hybrid Space: New Forms in Digital Architecture.* New York: Rizzoli.

Zulkifli, A.N., Alnagrat, A.J.A. and Mat, R.C. (2015). Development and evaluation of i-Brochure: A mobile augmented reality application. *Journal of Telecommunication, Electronic and Computer Engineering,* 8(10), 145–150.

Understanding Design Cognition in Computational and Generative Design

Design cognition is defined as the set of mental processes, strategies and knowledge areas employed whilst designing (Visser, 2004). To a general audience, the design process may appear mysterious or enigmatic. The term "black box" (Kurtoglu et al., 2008) nicely captures these characteristics, describing a situation which is so self-contained that its operations are difficult to understand. Researchers working in the field of design have, for many years, been focused on uncovering the mysteries and machinations of the design process. The tripartite goal of such research is to better understand the process, more effectively apply it in practice, and appropriately teach it to design students and novices alike. Traditionally, design researchers have used three main methods for this purpose: observation studies, interviews and protocol analysis (Ericsson and Simon, 1993; Gero and McNeill, 1998). In recent years, with advancements in neurophysiology, physical measures of eye-movement, body temperature, pulse rate and brainwaves have been gradually introduced and used to demystify design.

Design is not a linear process. Asimow (1962) suggests that the design process follows a spiralling path of analysis, synthesis and evaluation. This path typically includes proposition, testing, refinement, analysis and rejection stages, but they do not necessarily occur in that sequence. For example, in a study of Frank Gehry's approach, Boland et al. (2008) describe Gehry's design process as existing in a "liquid" state for a long period of time before eventually becoming "frozen" into a proposition for a building. During the liquid state, drawings and models are repeatedly made, tested and rejected as the design is cyclically refined until a "final" solution crystallises. In this way, Gehry and his colleagues explore and respond to different aspects of the design problem, alternatively shifting the focus from formal solutions to contextual problems, technological challenges and functional optimisation. For Gehry's team, this shift from problem definition and analysis to solution proposition and testing is often signalled by the decision to digitise a physical model for further development and refinement. While Gehry's forms and buildings may appear to be more complex than those of other architects, his team actually follows a relatively common cyclical design process that reflects their thoughts, actions and behaviours as they shift backwards and forwards between considering design problems and testing design solutions. Such examples, which can be seen in many architects' works, have been instrumental in the view that the design process formulates both a problem and ideas for a solution in parallel (Dorst and Cross, 2001).

One important factor influencing the design process is the medium or environment in which it is undertaken. Be it physical and sketch-based, or digital and CAD-based, the design medium and environment have a significant influence on designers' cognitive processes (Chen, 2001; Mitchell, 2003). As the emphasis of this book is on computational design, this chapter examines designers' cognitive behaviours in computational and generative design environments. This focus leads to three questions, which are the catalysts for this chapter and for Chapter 4.

- Does cognitive behaviour change in different computational design environments?
- What are the formal research methods used to examine design cognition and are they appropriate for different environments?
- What is the relationship between design thinking and computational thinking, and how do they interact with each other?

The answers to these three questions are gradually developed throughout this chapter and the next. However, rather than focusing solely on the results of past studies, this chapter interleaves discussion about research methods and the results developed by applying these methods. This structure is useful because understanding the mechanisms used to look inside the black box of design cognition assists us to critically appraise the results of this process. Furthermore, in Chapter 4 these methods are used to study multiple theories about design cognition which also answer some of these questions. In the following sections this chapter introduces and reviews key concepts related to design cognition before introducing methods, theories and some sample results.

3.1. Design cognition

The purpose of research into design cognition is to understand the process of design from a psychological perspective (Liikkanen, 2010). Cross (2001) defines design cognition as a cognitive science that studies problem-solving behaviour, including both problem finding and problem solving. To study design cognition, designers' thinking and processes are examined. In addition, improving our understanding of design creativity has always been one of the key goals of design research. Therefore, this chapter introduces and explores the relationship between design thinking, design problem finding/solving process and creativity.

3.1.1. Design thinking

Design thinking combines problem-finding and problem-solving behaviours (Cross, 2001). Design thinking is often described as comprising a set of six primary processes, with some involving multiple secondary variables. The first of these processes, *formulation*, involves defining requirements, intentions, abstractions, decisions and behaviours. The second, *synthesis*, is concerned with the generation of solutions. The *analysis* process determines the behaviours of solutions. *Evaluation* involves checking behaviours for viability and efficiency and then revisiting decision variables. The fifth process, *reformulation*, covers behavioural variables and intentions, and finally there is a *documentation* process (Gero, 1990). The nature of these processes is both exploratory and iterative, highlighting a form of agile thinking and responding. Through these processes, designers not only emphasise the generation of a solution or its alternatives, but also the constant redefinition of the problem. This redefinition is not included in conventional problem-solving practices, although current literature identifies it as an important process for achieving innovation.

Research on design cognition has been expanded over time, from the traditional focus of studying how designers think, work and collaborate (Cross, 2011; Lawson, 2005; Rowe, 1991), to understanding and applying such thinking to fields beyond design. Design thinking has been used to develop innovative solutions to a wide range of current and emerging problems facing industry and society (Brown and Katz, 2009; Liedtka et al., 2013; Martin, 2009; Moote, 2013). Design thinking has also been studied from a number of perspectives, particularly in relation to the roles of the design medium and environment (largely sketching in early studies) and the formation of formal cognitive models of design thinking. Protocol analysis, supplemented with different techniques, is the most common research method used for this purpose. For example, Suwa and Tversky (1996) explore the types of information architects extract from their sketches and the differences between expert and novice architects. The results of their study suggest that sketches are used to show perpetual relations and also non-visual functional relations. They also find that practising architects are more skilled at interpreting the function of a sketch than students. In a later study, Suwa and Tversky (1997) investigate why free-hand sketching is so useful for idea creation in the early phases of conceptual design development. They examine the information architects developed through the use of free-hand sketches while completing a set design task. Their results conclude that architects are able to interpret ("read-off") multiple visual and non-visual meanings and functional cues in sketches. With the ability to read-off functional relations and perceptions of depicted elements in a sketch, architects can pursue design thinking in greater depth. They are also able to shift focus more productively to consider alternative options, just as they shift focus to a new item, space or topic. More recently, Dinar et al. (2015) have reviewed a range of different studies and approaches, including comparisons of experts and novices, identifying and overcoming fixation, the role of analogies and the effectiveness of ideation methods. Their research suggests that there is a lack of standards related to this particular line of research, which may be due to the limits of cognitive models and theories for design thinking.

3.1.2. Design problems and design solutions

As discussed previously, design is not a linear process for finding solutions to an initial given task or requirement, it also involves redefining and reframing the design problems that have been provided (Asimow, 1962; Schön and Wiggins, 1992). During the design process, designers continue to redefine their design intentions, searching for alternative resolutions. This iterative process, which revisits both problems and solutions during the design process, should not simply be regarded as a cyclical series of events, because with each recursion in the process the parameters may have evolved and shifted. Previous studies show that the expert design process also involves a close interaction between the representations of problems and solutions (Cross, 2011). This co-evolution of the spaces of the design problem and design solution is one possible way to conceptualise the design process. Thus, instead of seeing design as a process of progressive refinement (that is, conceptual design leads to schematic design, then to developed design), design could be understood in terms of the way cognitive effort shifts between the consideration of problems and solutions (Dorst and Cross, 2001; Maher and Poon, 1996). As a result, design can be seen as a process that uses analysis, synthesis and evaluation as it shifts between the design problem and possible solutions (Asimow, 1962; Cross, 2011; Lawson, 1997). It is during this shifting process that designers formulate critical questions and explore answers, and thus the developing relationship between the "problem space" and the "solution space" is at the core of the co-evolutionary model of design. Maher and Poon (1996) and Dorst and Cross (2001) each use this model to suggest that co-evolution has a close correlation with the occurrence of design creativity.

Although it was developed for computational exploration (from an AI perspective), the model also describes a common design process. In the co-evolution model the problem space (P) and solution space (S) interact over time (t). Designers start by analysing the initial design requirements and formulating the design problem, P(t). While exploring possible design solutions S(t) for the problem P(t), new intentions are added into the problem space over time P(t+1). This is a core process for co-evolution in design, and particularly so when the solution does not satisfy a key requirement. By changing or adapting the requirements and intentions, a satisfactory problem and solution pair can be generated (Dorst and Cross, 2001; Maher and Poon, 1996).

Extending Maher and Poon's co-evolution model, Dorst and Cross (2001) propose a variation that illustrates the creative process from a behavioural perspective. In their model, the designers start from a design problem space P(t), and develop a partially structured problem space (P(t+1)Δ), which is then used to develop a partially structured solution space (S(t+1) Δ) of S(t). This process is repeated as the design progresses, with, as Maher and Tang (2003) suggest, the transition between design problem and solution occurring in cyclical iterations until a satisfactory solution is developed. Dorst and Cross (Cross and Cross, 1998; Dorst and Cross, 2001) further argue that this co-evolution process is vital for supporting the highest level of creative design.

3.1.3. Design creativity

Researchers usually approach creativity in design from four perspectives: creative processes, creative outcomes, creative individuals and interactions between creative individuals and the design context or environment (Said-Metwaly et al., 2017). In a survey of 152 creativity studies, Said-Metwaly et al. (2017) found that a focus on creative process is the most common in research that seeks to understand creativity (52.58% of studies), followed by a focus on creative person (28.87%), the outcome or product (14.43%) and finally interaction (4.12%). The most frequent of these, process exploration, has grown in strength over time as its results have been shown to be both replicable and generalisable (Rosenman and Gero, 1993; Suwa et al., 1999). The findings of some of this research are discussed in later sections, and it often identifies successful processes, for example, reformulation in creative thinking (Kan and Gero, 2008). In another example, Goldschmidt (2016) tests and discusses the hypothesis that frequent shifts between defocused and focused attention to stimuli in memory activation, which equate to divergent and convergent thinking, are a hallmark of creative thinking. The results of her research suggest that shifts between divergent and convergent thinking should be measured to better understand creative processes. The other major category of creativity research is the product-focused approach; this research evaluates the creativity exhibited in the design product (Runco and Pritzker, 1999; Torrance, 1966), with a variety of criteria used for assessing creativity, including novelty, surprise and value (Hayes, 1978; Maher, 2010; Nguyen and Shanks, 2009). Formal methods for evaluating creativity from the product perspective include Creative Product Semantic Scale (CPSS) (Besemer and O'Quin, 1993) and the Consensual Assessment Technique (CAT) (Amabile, 1982). Despite this research, and as originally noted by Hocevar (1981), defining and measuring creativity is a complicated process. Nevertheless, Sarkar and Chakrabarti (2008) argue that measuring creativity is essential to select innovative products and evaluate the degree of innovation taking place. The next two sections examine creativity research focused on the design product and then the design process.

Creativity and the design product

Researchers from psychology, social science, architecture, engineering and industrial design have all offered definitions of creativity. According to psychologists Runco and Pritzker (1999), the creativity of a product is characterised by a list of factors, including its "aesthetic appeal, novelty, quality, unexpectedness, uncommonness, peer-recognition, influence, intelligence, learning and popularity" (Sosa and Gero, 2004, p. 499). The measurement of the creative properties of an artefact is an important issue for researchers, designers and educators. Kaufman and Sternberg (2006), for example, claim that "creativity can be measured, at least in some degree" (p. 2). However, the evaluation of creativity can be subjective, and evaluation criteria and standards are not easily determined (Jordanous, 2011). As such, a first step in measuring creativity is establishing evaluation criteria. Cropley and Cropley (2005) propose a four-dimensional, hierarchical model for measuring the creativity embodied in a product. Their criteria are that it exhibits (i) relevance and effectiveness, (ii) novelty, (iii) elegance and (iv) generalisability. Rosenman and Gero (1993) argue that a creative product must also have a "richness of interpretation" (p. 111–2). This refers to the capacity to construe a design in multiple ways, and which to some extent represents potential for future development. Amabile (1983) argues that novelty and appropriateness should be the primary criteria for assessing a creative product. Reflecting the general lack of consensus on this topic, Ritchie (2007) and Oman and Tumer (2009) propose that novelty and quality should be the main criteria used to evaluate creative products. These two measures of product creativity have also been framed as novelty/originality and usefulness/value/utility in past research (Aldous, 2005; Boden, 2004; Cropley, 1999, Oman and Tumer, 2009; Ritchie, 2007; Weisberg, 1993; Wiggins, 2006). Sarkar and Chakrabarti (2008) conclude that most definitions of creativity include a consideration of both novelty and usefulness.

The creativity evident in a design product may also share some characteristics of the creative responses measured in psychology and cognitive science. Whilst many researchers use variations of the two criteria of novelty and usefulness, others argue that these two alone are insufficient to capture a third important emotional response, "surprise" (Brown, 2012; Bruner, 1962; Gero, 1996; Maher et al., 1996). A degree of unexpectedness is regarded as a measure of creativity in several fields. For example, Hayes (1978) and Nguyen and Shanks (2009) use three criteria – value, novelty and surprisingness – to assess creativity. Maher (2010) uses value, novelty and unexpectedness, and proposes a set of formal methods to measure surprise. Brown (2012) further proposes a framework to consider the types of situations where designers or evaluators might be surprised.

In addition to the evaluation criteria, some formal methods have been developed for assessing creative products. One of the most well-known methods is the Creative Product Semantic Scale (CPSS) (Besemer and O'Quin, 1993) and its evaluation criteria are as follows. "Product novelty" relates to whether or not an object is original and surprising. "Product resolution" describes the value, logic, use or demonstrable application of the object. "Product elaboration and synthesis" is concerned with the extent to which an object is refined, elegant or well-crafted. An alternative to CPSS is the Consensual Assessment Technique (CAT) (Amabile, 1982), which uses novelty and value as the criteria for evaluating creativity in a product. The CAT approach assesses creativity using expert evaluation by judges. Both CAT and CPSS are ultimately founded on structured, subjective assessment, often supported by statistical analysis.

Sarkar and Chakrabarti (2011) propose a method for assessing the degree of creativity of design products based on definitions of novelty and usefulness. Their method employs both the Function-Behaviour-Structure (FBS) (Gero, 1990) and SAPPhIRE models (Chakrabarti et al., 2005). The FBS model is used to test if a design process is novel or not, and then the SAPPhIRE model is used to assess if the design product is novel. The FBS model is more commonly used to identify the three types of reformulation processes that potentially capture innovative or creative aspects of designing (Kan and Gero, 2008; Kan and Gero, 2009). In the work of Sarkar and Chakrabarti (2011) a source of comparison between the process and the product is provided. Jagtap (2018) further develops this method by proposing various modifications that support evaluation creativity by benchmarking against the collective, intuitive assessment of experienced designers. The goal is to ensure a more consistent standard and a higher degree of objectivity.

Creativity and the design process

In the past, researchers have studied the design processes involved in the production of creative products and recognised that evaluating the creativity implicit in the processes is a complex issue (Lawson, 1997). Fundamental to improving our understanding of the role of creativity in designing is determining if there are specific processes that lead to creative outcomes. These "creative processes" have been defined in a variety of ways. One common definition is that "creativity occurs through a process by which an agent uses its ability to generate ideas, solutions or products that are novel and valuable" (Sarkar and Chakrabarti, 2011, p. 349). Sarkar and Chakrabarti's definition of creativity reflects the common view that a creative process is one that can produce a creative outcome. However, it is possible that a creative process will not produce a creative product, and conversely, a creative product could arise out of a relatively mundane or formulaic approach. The two, process and product, do not necessarily exist in a fixed or causal relationship. Nevertheless, as the previous sections note, there is growing evidence that certain cognitive processes and strategies are more likely to produce creative outcomes.

The stages or components in a creative design process have been defined in various ways. For example, Wallas (1926) argues that there are four stages: preparation, incubation, illumination and verification. These four have also been used in past research (Dewett, 2003; Kristensen, 2004; Rastogi and Sharma, 2010) and been shown to provide a robust and useful framework (Runco, 2004, p. 665). For measuring a creative design process, however, Guilford (1975) suggests that divergent thinking is the most important component that must be captured. Divergent thinking tests have been widely used for measuring creative processes or creativity-relevant skills (Kaufman and Sternberg, 2006). Another approach is to measure the extent to which a co-evolution process is evident (Yu et al., 2015). This approach examines the way cognitive effort shifts between the consideration of problems and solutions (Dorst and Cross, 2001; Maher and Poon, 1996). The rationale for this method is drawn from the view that a creative design process requires particular combinations of problem-finding, idea-finding and problem-solving processes (Osborn, 1963; Parnes, 1981).

To optimise the assessment of a creative process, and to tailor it for a different context, multiple variations or combinations of these methods have been proposed. The results of these studies have deepened our understanding of the creativity of the design process. For example, Hasirci and Demirkan (2007) use observation, protocol analysis and rating scales to assess creativity during the design process. Their research shows that creative design products and processes are highly correlated. Toh and Miller (2015) evaluate creativity from the perspective of concept selection in a design teamwork context. The results of their study suggest that creative concept selection may be related to discussions about the decomposition of generated ideas. D'Souza and Dastmalchi (2016) evaluate creative design processes by applying the CAT method (Amabile, 1982) and suggest that design processes do not follow patterns of linear periods of incubation, followed by creative leaps. Instead, the expertise and background of designers are critical for creative design idea generation. Other creative process measurement methods include the Wallach-Kogan Creativity Test (WKCT) (Wallach and Kogan, 1965) and the Structure of the Intellect Divergent Production Test (SOI) (Guilford, 1967).

As the previous section notes in respect of creative products, a creative process can be one that results in "effective surprise" (Bruner, 1962, p. 3). Within the specific context of design cognition, Gero (2000) builds on this view, defining process creativity as the "activity that occurs when one or more new variables is introduced into the design" (p. 187). This way of viewing creativity in the design process potentially offers several further ways of formally studying cognitive creativity.

3.2. Formal approaches to studying design cognition

3.2.1. Protocol analysis

Protocol analysis is a method for turning qualitative physical actions and verbal and gestural utterances into quantifiable data (Ericsson and Simon, 1993; Gero and McNeill, 1998). It has been used extensively in design research to develop an understanding of design cognition (Atman et al., 1999; Kan and Gero, 2008; Suwa and Tversky, 1997). According to Akin (1986), a protocol is the record of the behaviours of designers, made using sketches, notes, videos or audio. After collecting the protocol data, a coding scheme is applied to it, to categorise and collate the information in a quantifiable way. This enables a detailed study of the design process in the chosen design environments or settings. As Gero and Tang (2001) note, protocol analysis has become the prevailing experimental technique for exploring design cognition and the design process.

In a conventional application of protocol analysis, concurrent or retrospective protocol collection methods are applied in design experiments (Dorst and Dijkhuis, 1995; Ericsson and Simon, 1993). The concurrent protocol involves participants verbalising their thoughts while working on a specific task, which is also called the "think-aloud" method. A retrospective protocol involves participants reflecting on what they were thinking while designing, a process that occurs as soon as they have finished the design task. These two protocol collection methods have their distinct advantages and limitations. For instance, Kuusela and Pallab (2000) argue that concurrent protocols are more suitable for examining the design process and can generate larger numbers of segments, while retrospective protocols are more suitable for examining design outcomes. Another view is offered by Gero and Tang (2001), who show that concurrent and retrospective protocols can lead to very similar outcomes when exploring designers' intentions during the design process. However they also conclude that concurrent protocols are the more efficient and applicable method for understanding design. On the other hand, retrospective protocols are commonly believed to be less intrusive, because the "think-aloud" method may be a distraction from the act of designing.

In addition to considerations of concurrent versus retrospective variations, there are two approaches to protocol analysis: process-oriented and content-oriented (Dorst and Dijkhuis, 1995). The process-oriented approach seeks to encapsulate a designer's actions, such as developing design plans, goals and strategies. The content-oriented approach focuses on indicators of cognitive problem-solving, looking at what designers see, and the knowledge they use to achieve a design outcome (Suwa and Tversky, 1997).

Protocol analysis procedure

There are several common procedures for developing a protocol analysis study (Ericsson and Simon, 1993), most of which include variations of the following components.

1. Developing a hypothesis or determining a goal for the study.
2. Selecting or developing a coding scheme that is aligned to the initial goal or hypothesis.
3. Experimental design and participant recruitment.
4. Conducting the experiments.
5. Transcribing and segmenting the materials generated in the experiments.
6. Encoding the protocol data.
7. Quantitative and qualitative comparison of protocol data.
8. Framing results in terms of the initial hypothesis or goal.

In addition to these components, there are also several alternative methods for segmenting data, such as dividing it by fixed time duration, by individual sentence or by the meaning of the protocol. Several methods exist for confirming the initial coding of the protocol data (arbitration processes) and there are also statistical methods for measuring the reliability of the result.

Protocol studies in design research

Protocol analysis has been used in the past to study designers' cognitive behaviours in different design environments (Bilda and Demirkan,

2003; Bilda et al., 2006; Kan and Gero, 2009; Kan et al., 2011; Kim and Maher, 2005; Tang et al., 2009; Tang et al., 2011). To capture the breadth and depth of protocol studies in design research, a literature review was conducted for this chapter using *Scopus*. The following parameters were used. Keywords were "protocol analysis" and "design" ($n = 1,539$) and subject areas restricted to "Engineering", "Social Science" and "Art and Humanities". The document type was limited to "article", and only publications between 1996 to 2020 were selected ($n = 255$). Using journal classifications, the scope was further limited to 19 design, architecture, engineering and education journals. At the completion of the search, the list of relevant journal articles ($n = 110$) were identified for review.

Table 3.1 shows that the majority of protocol studies were published in the years between 2016 and 2020, with the overall number indexed in *Scopus* increasing across each time period to the present. Most of the protocol research produced between 2000 and 2005 is focused on designers' cognitive behaviour in sketching environments, which is a traditional and essential skill for designers. During the 2006–2010 period, emerging digital design tools brought new opportunities as well as challenges for designers. Researchers during this period explored whether or not new digital technologies can assist designers' cognitive processes. As a result, the protocol studies during this period were often focused on designers' behaviour in digital environments, such as Computer-Aided Design (CAD) modelling, digital sketching, Tangible User Interfaces and haptic interfaces. From 2011, with the emergence of next generation CAD tools, including parametric and generative design, researchers started to study designers' behaviour in these new environments. From this literature review, the large and growing number of protocol studies confirms that it is one of the most frequently applied and reliable techniques in design studies.

Overview of coding schemes applied in protocol studies

In any protocol study, the coding scheme is one of the most important elements. It is used for encoding and analysis of the data collected from the design experiment. The earliest coding scheme in cognitive studies was proposed by Eastman, who used design units, constraints and manipulations to encode protocols and explore a behaviour graph of the design process (Eastman, 1970). Since then, researchers have continued to develop a variety of coding schemes to address specific research questions.

The literature review identifies two significant coding schemes. The first is focused on the design-action categories originally proposed by Suwa and Tversky (1997), which represents a content-oriented approach. The second is based on Gero's (1990) FBS model, which seeks to provide

Table 3.1. Journal articles (1996–2020) adopting protocol analysis as the main method (Source: Scopus).

Year	Articles	Number
1996–2000	Atman and Bursic, 1996; Atman and Bursic, 1998; Frankenberger and Auer, 1997; Galle and Béla Kovács, 1996; Mc Neill et al., 1998; Suwa et al., 1998; Suwa and Tversky, 1997; Welch, 1998; Welch et al., 2000	10
2001–2005	Akin and Moustapha, 2004; Atman et al., 2005; Bilda and Demirkan, 2003; Chakrabarti et al., 2004; Chan, 2001; Davis et al., 2002; Gero and Tang, 2001; Ho, 2001; Kavakli, 2001; Kavakli and Gero, 2002; Meniru et al., 2003; Seitamaa-Hakkarainen and Hakkarainen, 2001; Sim and Duffy, 2003; Stempfle and Badke-Schaub, 2002; Taura et al., 2002; Wu and Duffy, 2004	17
2006–2010	Ahmed, 2007; Al-Sayed et al., 2010; Almendra and Christiaans, 2009; Azevedo and Jacobson, 2008; Bilda and Gero, 2007; Bilda et al., 2006; Cardella et al., 2006; Coley et al., 2007; Goldschmidt et al., 2010; Houseman et al., 2008; Ibrahim and Pour Rahimian, 2010; Jin and Benami, 2010; Jin and Chusilp, 2006; Kim et al., 2007; Kim and Maher, 2008; Lemons et al., 2010; Liikkanen and Perttula, 2009; Menezes and Lawson, 2006; Strickfaden and Heylighen, 2010; Wang et al., 2010	21
2011–2015	Bertoni, 2013; Cash et al., 2015; Chai and Xiao, 2012; Chandrasekera et al., 2013; Cheong et al., 2014; Deken et al., 2012; Ensici et al., 2013; Eseryel et al., 2013; Huang et al., 2012; Jagtap et al., 2014; Kelley et al., 2015; Kim and Kim, 2015; Kim et al., 2011; Leblebici-Başar and Altarriba, 2013; J. Lee et al., 2014; Liedtka et al., 2013; López-Mesa et al., 2011; Mentzer et al., 2015; Mohamed Khaidzir and Lawson, 2013; Nikander et al., 2014; Rahimian and Ibrahim, 2011; Tang et al., 2011; Vallet et al., 2013; Wang et al., 2013a; Wang et al., 2013b; Yu et al., 2013; Yu et al., 2014	28
2016–2020	Atman, 2019; Blom and Bogaers, 2020; Brösamle and Hölscher, 2018; Casakin, 2019; Cash and Kreye, 2018; Cash and Maier, 2016; Chu et al., 2017; Cramer-Petersen et al., 2019; D'Souza and Dastmalchi, 2016; Dixon and Bucknor, 2019; Ferreira et al., 2016; Grubbs et al., 2018; Hay et al., 2017a, 2017b; Hu et al., 2019; Kan and Gero, 2018; Kannengiesser and Gero, 2017; Kelley and Sung, 2017; Lee et al., 2018; Lee et al., 2019; Lee et al., 2016; Mao et al., 2020; Nguyen et al., 2019; Önal and Turgut, 2017; Sauder and Jin, 2016; Shih et al., 2017; Sung and Kelley, 2019; Tedjosaputro et al., 2018; Tracey and Hutchinson, 2019; Wells et al., 2016; Yang and Lee, 2020; Yang et al., 2020; Yu and Gero, 2016; Yu et al., 2018	34

the basis for a universal coding scheme applicable to the broader design domain. FBS is a process-oriented approach that focuses on a designer's intentions. It was developed in response to the realisation in the early years of protocol research that there were few coding schemes that were repeatedly used, and therefore comparing and validating experimental results was difficult (Kan and Gero, 2009). The next two sections examine Suwa's and Gero's schemes in more detail, before considering some alternatives.

Protocol studies using Suwa's coding scheme

First established by Suwa and Tversky (1997), and later revised by Suwa et al. (1998), Suwa's coding scheme is widely used in cognitive design research and especially in studies about the design process and free-hand sketching. Tang and Gero also conducted a series of design research studies based on this coding scheme (Gero and Tang, 2001; Gero and Tang, 1999; Tang and Gero, 2000; Tang and Gero, 2001). This coding scheme uses four categories to encode and distinguish design actions: physical, perceptual, functional and conceptual (Suwa et al., 1998). By using these classifications researchers are able to describe the inter-relationships between these design actions and the designer's thoughts.

Since being proposed in the 1990s, many cognitive design studies have used or adapted Suwa's coding scheme. For instance, it has been used to explore the interactions between sketching and goal-setting (Gero and Tang, 2001), and to suggest that visual reasoning is more significant in the design process than was previously thought (Bilda and Demirkan, 2003). Another study based on Suwa et al.'s revised coding scheme is Kim and Maher's comparison of Graphical User Interface (GUI) and Tangible User Interface (TUI) environments in collaborative design environments. With the support of this coding scheme, their research was able to show that the use of TUI supports designers' spatial cognition (Kim, 2006; Kim and Maher, 2005; Kim and Maher, 2008). With the emergence of parametric design environments (PDEs), researchers started to consider this environment using protocol analysis. For example, Lee et al. (2014) present a pilot study using protocol analysis and Suwa's coding scheme to evaluate creativity in PDEs. The results of that study identify specific conditions in PDEs that can potentially enhance design creativity. Beyond the focus on design environments and representations, this coding scheme has also been used to propose a refined model of the co-evolution of problem and solution spaces in design (Dorst and Cross, 2001). That study supports Schön's (1983) argument that insight-driven problem reframing is crucial to the creative design process.

Protocol studies using the FBS coding scheme

The coding scheme based on the FBS model (Gero, 1990) adopts a process-oriented approach that focuses on encoding the intentions of designers. It contains five categories: function (F), expected behaviour (Be), behaviour derived from structure (Bs), structure (S) and description (D). It was first established by Kan and Gero (2009) before being applied in a growing numbers of studies of design collaboration. An example is found in Kan and Gero's own research (Kan and Gero, 2009), which used the FBS coding scheme to study different forms of collaborative design activity, presenting the results of alternative expressions in formulation and reformulation processes. Later, they described an attempt to explore and use quantitative research tools to examine design protocols collected in a collaborative virtual environment. Their results show that, compared to face-to-face design collaboration, the 3D virtual environment slows design activities and has a tendency to favour certain activities during distant collaboration (Kan and Gero, 2010). The following year, Kan and Gero used the FBS coding scheme to evaluate the learning process of a design team. In that particular study, "linkographs" were also used to examine team interaction and individual design processes (Kan et al., 2011). Linkography is a technique used in protocol analysis to assess the design productivity of an individual designer (Kan and Gero, 2008). It involves looking at each design move as a step/act or operation that ultimately changes the design situation. A linkograph is developed by assessing the relationships between each act to form links. This tracing of associations provides a graphical representation of the design session, and the links then form patterns which display the structure of the design reasoning (Kan and Gero, 2008). The FBS ontology has also been used to study designers across multiple disciplines, in which both commonalities of design and unique characteristics within disciplines have been identified (Jiang, 2012; Jiang et al., 2014).

A common use of the FBS coding scheme is to compare design processes in different design environments. There are two reasons for this. First, the FBS model is capable of capturing designers' high-level thinking, and second, the FBS model provides a universal coding scheme and therefore does not contain very detailed design actions. This means it can be easily adapted to suit different design environments, contexts and disciplines. For instance, Tang et al. (2009) used the FBS coding scheme to compare free-hand sketching and digital sketching, and then argue that the two design environments are similar in terms of support for design efficiency, process and content. In a later study, they propose that the design processes in these two environments are not statistically different in terms of the distributions and transitions of specific processes (Tang et al., 2011).

More recently, an augmented ontological FBS model has been developed (Gero and Kan, 2016) to generate quantitative descriptions of cognitive behaviour in terms of design creativity by adding two variables, "new" and "surprising", to capture and formally study design creativity. The findings of this research support the claim that novelty and surprise are key elements for design creativity and that these can be measured empirically. This study also indicates that it is possible to apply the FBS coding scheme to recode existing protocols, to develop results for understanding and theorising creative behaviour without having to conduct new experiments.

Other coding schemes and summaries

In addition to the two main coding schemes, there are several other schemes available. For example, seeing–imaging–drawing (S-I-D) and seeing–seeing as (S-SA) (McKim, 1980); seeing (S) and seeing as (SA) (Goldschmidt, 1989, 1991; Schön and Wiggins, 1992); novel design decision (NDD) (Akin and Lin, 1995); combinations of S-I-D, SA- seeing that (ST) and total-detail (T-D) (Won, 2001) and analysis-synthesis-evaluation (Gero and McNeill, 1998). These coding schemes were generally developed and used in early cognitive studies, providing a variety of ways to understand design processes. However, none of these are as broadly applicable and systematically adopted and enhanced as those of Suwa and Gero. Nevertheless, customised coding schemes continue to emerge to satisfy specific needs. For example, Chien and Yeh (2012) explore "unexpected outcomes" in Parametric Design Environments (PDEs) using a customised coding scheme to accommodate "problem structure-searching-feeling-ideating" during parametric design. In another example, Lee et al. (2019) demonstrate a dual-coding scheme for protocol analysis, one coding drawn from Suwa et al. (1998) and the other adapted from spatial language theory. They use this dual-coding to examine cultural and linguistic differences in design cognition.

In summary, Suwa's and Gero's coding schemes are still the most widely used in cognitive design studies to date. In comparison with Suwa's coding scheme, Gero's FBS scheme is structured at a higher level due to its process-oriented approach, focusing on the intentions of the designer. In contrast, Suwa's coding scheme contains specific design actions, particularly suitable in sketching environments, although it can also be extended and adopted for other design environments including computational environments.

3.2.2. Biometric approaches to studying design cognition

In recent years, with advancements in neuroscience, several new techniques have been developed or adapted for studying designers' cognitive behaviour. In most cases they use biometric or neurophysiological evidence to examine the design process. The remainder of this section introduces design cognition studies using eye-tracking and electroencephalogram (EEG), two recent biometric approaches to cognitive design research.

Eye tracking technology

Eye-tracking systems measure eye position, eye movement and pupil size to define the direction and duration of a person's gaze (Gonzalez-Sanchez et al., 2017). Early work in eye-tracking technology can be traced to Dodge and Cline (1901), who developed the first precise eye-tracking technique, which used light reflected from the cornea. Since the 1970s there has been a rapid growth of eye-tracking applications, often driven by research in advertising, marketing, psychology, neuroscience, design cognition and user interface design. There are two main types of eye-tracking devices: screen-based eye-trackers and mobile eye-tracking glasses. The data collected using an eye-tracking system typically includes heat maps, pupil size, viewing sequence and fixation duration. Examples of two popular eye-tracking systems in current use are *Tobbi Studio* and *Gaze Point*.

Studies of visual attention on static pictures using eye-tracking

The neurophysiological or biometric impact of photographs, artworks, sculptures and advertisements can all be measured using viewer eye-movement. Multiple past studies have established the relationship between eye movement and perception of static pictures. For example, Arnheim (1974) found that people tend to read pictures from left to right. Torralba et al. (2006) state that context information plays an important role in object detection and observation, and some parts of a scene can attract more attention than others. Jacob and Karn (2003) studied eye movements while utilising a user interface within human-computer dialogue. Their study shows that the number of fixation zones (Fixation Frequency) in each area of interest can serve as a measure of the importance of the information content in that area, in comparison to other areas. Eye-fixation times can be measured in various pre-defined parts of figures, with the centre of gravity treated as an attractor and the edges and corners reflecting the area of influence (Kaufman and Richard, 1969). The viewer's interpretation of the figure does not affect eye movement; instead only "physical attributes" influence eye movements (Gould and Peeples, 1970). Using eye-movement data, the 3M company developed visual attention simulation (VAS) software, which can simulate the way viewers will inspect static pictures. Studies show that this software provides a reasonable facsimile of human viewing behaviours under certain conditions (Auffrey and Hildebrandt, 2014).

Studies of visual attention on 3D spaces using eye-tracking

While the relationship between eye movement and perception in artworks has been investigated, there has been very little research into the role of eye movement in the perception of three-dimensional spaces and architectural forms. One of the few studies with a focus on architecture, from Weber, Choi and Stark (2002), collected eye-tracking data from participants who were asked to look at three-dimensional models or photographs of models depicting an architectural space. Their results show that, with no model in a figure, the attention of the participants moved to the centre of the image. With the model in place, the foreground was a common location for initial fixations, and the eye typically did not scan the edges of interior spaces or rectilinear contours. Furthermore, objects on the left attract more attention than those on the right. In essence, their research compared different arrangements of objects in a space, rather than different methods of representing the same spatial configuration. Their findings echo the results of some recent research (see Chapter 4), which supports the validity of using eye-tracking as a method in design cognition studies. Weber et al.'s (2002) study further reveals that fixations do not vary significantly when viewing a physical model compared with a photograph of the model, with the exception of the foreground, which attracted greater attention in the physical model. Additionally, their results also suggest that there can be significant differences between the fixations and saccades (simultaneous movements of both eyes between fixations) of architects and non-architects. These observations would not have been possible without the use of technology.

Eye-tracking technology has also been used for measuring the emotional reactions of viewers to interior spaces. Tuszyńska-Bogucka et al. (2019) conducted an eye-tracking experiment to measure respondents' reactions while looking at visualisations of interiors, with the aim of verifying whether certain parameters of an interior are related to emotional reactions in terms of positive stimulation and the sense of security or comfort. The results show that the varying spatial and colour arrangements presented in interior visualisations can provoke different emotional responses, supported by pupil reaction parameters, as measured by eye-tracking devices. They conclude that architectural space can have a diverse emotional significance and impact on an individual's emotional state. This has added a further rich layer of information to studies of design cognition.

Exploring design thinking using eye-tracking

Eye movement can also reflect human thought processes, because a person's thoughts may lead them to look into a certain location, or saccade between fixation points (Yarbus, 1967). This correlation of eye-movement and design thinking is one possible way to examine the mystery of the "black box" during a design process. One research approach is to use a combined method of protocol analysis and eye-tracking, and the validity of this method was tested by Guan et al. (2006). In the study, they conducted a design experiment in which participants' eye movements were recorded and subsequently analysed, while performing think-aloud to allow correlational analysis of the two data sets. Results of their study show that the recounting of what went on in the exercise was consistent with the sequence of objects viewed and in the same order. They also found that the differing level of complexity did not interfere with the validity of retrospective think-aloud. The exploratory information generated from retrospective think-aloud can also indicate how the information was processed by the user, and what particular strategies were used.

Design representation is another important factor for both design thinking and design communication (Self et al., 2014). With the development and widespread adoption of computational modelling, digital design representations now dominate many design processes, assisting designers in both "off-loading" cognitive load and providing the possibility of interacting with other external representations (Schön and Wiggins, 1992). The role of representation in the architectural design process has been explored by Park et al. (2019) using eye-tracking technology. They conducted a design experiment in which eye-tracking data were collected from participants viewing six pairs of photographs and line drawings and analysed to understand how representations affect people's perceptions of architectural scenes. The results of their study suggest that the line drawing variations can both attract and deflect a participant's visual attention. For example, a line drawing may increase visual attention if its elements are clearer or more emphasised than those in a photograph of the same image.

To understand the underlying relationships between visual behaviour and cognition, past studies have suggested that cognitive transfer is required to read a two-dimensional drawing as opposed to a three-dimensional mental representation (Lohmeyer et al., 2014). Designers also tend to think sequentially and visually to understand details (Lohmeyer et al., 2014) and level of experience or expertise also has an impact on visual behaviour. Cao et al. (2018) compare the different behaviours of novice and advanced design students by analysing their eye movements during conceptual idea generation using analogical reasoning. This approach also used solution evaluation and retrospective interviews to support the analysis. Their study finds that improvements can be made to beginner students' cognitive behaviours, to both help them to grow their capacity for design thinking and improve their design outcomes. Cao et al.'s (2018) research design is an example of contextualising eye-tracking technology for cognitive design research.

Electroencephalogram (EEG)

EEG is a tool that is increasingly being applied to measure designers' biometric responses during the design process. EEG is used for analysing brain activities and human behaviours using the frequencies of brain signals (Kumar and Bhuvaneswari, 2012). The application of EEG in design studies emerged during the last decade. It has been used to measure a designer's brain waves, showing that designers spend more effort engaged in visual thinking in the solution-generation phase than in the solution-evaluation phase (Nguyen and Zeng, 2010). In a more recent study, the same approach was used to investigate the relationship between the designer's mental effort and stress levels, where the participants' design activities, body movements, brain signals and heart rates were recorded during a design task (Nguyen and Zeng, 2014). Results suggest that physiological parameters may reflect stress, but effort – such as body movement or eye-blink frequency, and the relationship between mental stress and effort – may differ depending on the design task, participants' personality or expertise level. Furthermore, using the transient microstate percentage of EEG signals (Nguyen et al., 2015), the perceived cognitive difficulty of a design problem can potentially be measured.

In the latest developments, EEG has been combined with other biometric approaches including eye-tracking and gesture analysis – based on signals generated from galvanic skin response (GSR), electrocardiograms (ECG) and EEG – to more comprehensively study designers' processes (Nguyen, 2017). The gesture analysis measures designers' mental effort, concentration and fatigue during the conceptual design phase. Nguyen's (2017) results show that mental effort is related to the spatial-temporal patterns of EEG, which can be used for studying design cognition. For example, end of task phenomena, measured by extended fatigue level, was identified during the design experiment. Fatigue and mental effort conform to a typical "capacity" model and shifts in concentration might be indicators of creativity. These emerging topics in design cognition have been further examined (Nguyen et al., 2018), supported by empirical evidence produced using EEG during design experiments. The results of Nguyen et al.'s (2018) study suggest that high levels of mental effort occur at the beginning and end of design tasks, and that this effort correlates negatively with metrics of fatigue. This methodological approach has enabled issues, such as the designer's mental effort, fatigue and concentration during the design process, to be quantified, which has not been possible using previous design cognition methods.

EEG has also been used for exploring different design activities during the design process. For example, Liu et al. (2016) demonstrate that results in specific EEG bands correlate to design activities, such as problem solving and evaluation. This provides a unique opportunity for research into design cognition, as well as recognition and exploration of different activity patterns during the design process. Another study by the same team focuses on designers' activities, divergent thinking, convergent thinking and mental workload. The EEG-based results identify a relationship between the design process and different parts of the designer's brain activities (Liu et al., 2018).

As shown in these examples, biometric approaches to cognitive design studies can provide us with empirical evidence to enhance and complement our current knowledge about design cognition. Using biometric measurements of designers during the design process, data related to physiological responses can be obtained to provide empirical evidence. Analysis and correlation can then be conducted to explore the relationship between the designer's physiological responses and the specific focuses of the design research. In contrast to these biometric approaches, traditional research methods such as protocol analysis rely heavily on designers' personal accounts of their thoughts and actions, which can be distorted through a subjective lens. What designers say and (sometimes) think they are doing while designing may not always be an accurate reflection. On the other hand, protocol studies have a long tradition in cognitive design research, and their results have been validated, debated, refined and documented over a significant body of work. They also provide an important theoretical foundation for benchmarking and contextualising new findings. Having different research tools allows design scholars to choose the most suitable one for the intended research problem, which is more important than promoting a single approach for all problems.

3.3. Design cognition in the computational design environment

With the increasing adoption of CAD technologies across the AEC sector, designers' approaches are undergoing significant changes due to the emergence of new computational design environments. In many of the computational design environments, designers' cognitive efforts must not only contend with the design process itself, but also with the use of computational design tools (Yu et al., 2015). This imposes another cognitive layer beyond traditional design thinking, which is known as 'computational thinking'. This section introduces computational thinking and reviews recent studies on design cognition in computational design environments. In particular, it looks into the impact of generative and parametric design on designers' behaviours.

3.3.1. Computational thinking and design thinking

Interest in computational thinking has grown over the last decade as a separate and significant area of research for software development. A common definition of computational thinking emphasises the importance of *formulating*, *organising*, *automating*, *identifying* and *generalising* (Wing, 2008). These terms are defined as follows.

1. Formulating problem parameters in such a way that computational methods can be used to solve them.
2. Organising data in such a way that it can be analysed, represented and abstracted for use in simulations or models.
3. Automating processes in accordance with algorithmic logic and iterative decomposition and recomposition of parameters.
4. Identifying and implementing possible solutions in accordance with given parameters (efficiency, effectiveness, resourcefulness or factor parsimony).
5. Generalising a problem-solving process for application to other situations.

Computational thinking can be distinguished from mathematical thinking (or engineering, or scientific thinking) by virtue of having a capacity for symbolic abstraction that is sufficiently general to be applicable in other disciplines (Wing, 2008). As evidence of this, it has been extended to serve as a general cognitive model for the development of all products and processes and for understanding abstract concepts (Bell, 2005; diSessa, 2005; Kafura and Tatar, 2011; Moursund, 2006; Resnick et al., 2009; Stonedahl et al., 2009). There are, however, relatively few studies that directly address computational thinking in design. Nevertheless we can look at the role and impact of computational thinking in design through the application of computational design environments where this mode of thinking is reinforced. Ekströmer (2019), for example, suggests that the use of CAD tools in the design ideation process can assist with design thinking. Furthermore, it has been claimed that computational design tools are able to support higher levels of creative agency and even become a partner in the design process (Mothersill and Bove, 2019).

3.3.2. Design cognition in the computational environment

The combination of computational thinking and design thinking has been applied to an increasing number of domains in industry and society. Collectively, they offer a powerful conceptual paradigm for reasoning and problem-solving, providing innovative approaches for understanding and changing our world. Importantly, the formal processes defined previously in this chapter for design thinking and computational thinking have similar goals. The primary difference is that design thinking does not specify that the processes are carried out using computation, while computational thinking does. Furthermore, computational thinking has an additional process—that of generalising and transferring the processes developed. This gives computational thinking a formalised, externalised output that transcends the particular problem being solved and which increases its application in all domains, including design. Combining these two approaches has significant potential to create new and more powerful conceptual paradigms to develop innovative strategies for all aspects of contemporary industry and society. Since the present book is concerned with design, the focus of the remainder of this chapter is on reviewing and understanding design cognition in a computational environment. This involves the application of the key approaches and methods introduced and reviewed previously (Section 3.2) but more exclusively in the computational design environment.

Investigating design thinking in computational environments

The impact of computational design environments on design thinking has been explored by researchers from various perspectives. Some of these studies argue that computational design tools restrict design thinking in certain situations. For example, Stones and Cassidy (2007) studied 96 novice graphic designers, comparing paper-based tools and computer-based ones, and found that computer-based tools may limit the scope of ideas being generated. Their study concludes that designers are less likely to achieve an integrated design in a computational design environment. This result may be due to the lack of technical skills of the participants, or, more interestingly, they may not be able to "see" the full possibility of the relationships between forms when using the computer. Another study suggests that a "bottom-up" approach, as demonstrated in many computational tools, supports improved intuitive responses to evaluating design alternatives (Mothersill and Bove Jr, 2018). The difference between these findings could be shaped by the diverse backgrounds and preferences of individual designers in each study. Equally, this disagreement emphasises that different design tools have different capacities, and designers should be able to choose "the right tool for the right task".

Cognitive behaviour and computational design tools

A common goal of many past studies of computational design tools has been developing new tools to support designers and design environments. Focusing on idea formulation and generation, some studies indicate that designers prefer to directly manipulate geometry and to change between different "views", both of which indicate the need for particular tools or functions in software (Guidera and MacPherson, 2008). Do and Gross (2001) discuss the requirements for CAD tools to support design thinking, noting as a starting point that free-hand sketching is effective for conceptual design in architecture. They argue that the process of drawing diagrams – bubble diagrams, sight-line diagrams and diagrams of abstractions and representations for more specific elements – should be better supported by digital design tools. Further features for computational design tools include: capacity for freehand drawing; means for maintaining spatial relations between elements as a drawing is altered; capturing emergent patterns in a diagram; automated transformations between diagrams; identification of similarities and differences between diagrams; and capacity to use varying levels of abstraction and detail in a diagram (Do and Gross, 2001). Other considerations for future tools include the need for better representation of the tacit aspects of design knowledge, as well as a change in focus from tools that just support modelling to those that support designers' needs (Bernal et al., 2015).

3.3.3. Design cognition in the Parametric Design Environment (PDE)

As described previously in Chapter 2, parametric design is an emerging computational design approach in the AEC sector, where a dynamic and algorithmic rule-based design process is supported to enable multiple design solutions to be developed in parallel. Karle and Kelly (2011, p. 109) define parametric thinking as a method for connecting "tangible and intangible systems into a design proposal" and then formalising "relationships between properties within a system." It is a cognitive approach that requires designers "to start with the design parameters and not preconceived or predetermined design solutions" (p. 109). The theorised changes of designers' activities in a Parametric Design Environment (PDE), compared to those in a Traditional Design Environment (TDE), are summarised as follows.

- *Designers design rules and define their logical relationships rather than only modelling geometries*
  One of the significant differences between parametric and traditional design is that rule-sets comprise the basic design procedures in a PDE (Abdelhameed, 2009). During the process of creating models, designers set variations, identify data flows, adjust parameter values and revise rules. They are not only thinking about the particular building design but also the rules required to achieve the building design. Additionally, through the control of logical relationships in the algorithm, more possibilities for design solutions are generated (Hernandez, 2006; Karle and Kelly, 2011).
- *Designers may freely alter all the steps of their process*
  In a parametric design process, designers are able to revisit any previous step or action to change parameters or revise rules and thereby alter the design. This level of flexibility allows the design process to remain open-ended until the right solution is derived. A further advantage is that each stage in a parametric design process is effectively captured in the algorithm and its application, providing a level of documentation suitable for knowledge capture purposes.

- *Large numbers of design alternatives can be developed in parallel*
  Designers in TDEs typically consider only a small number of alternatives because of time limitations (Woodbury and Burrow, 2006). In a TDE, as Akın (2001) argues, design solutions are not optimal, but satisfactory. In a PDE, once the rules are set, large numbers of design alternatives may be easily generated. This process provides a variety of possibilities, as well as widening the designers' thinking. The alternatives can also be developed in parallel, so designers do not need to predetermine design solutions at an early stage (Hernandez, 2006; Holland, 2011; Karle and Kelly, 2011). Parametric design allows this level of intelligence to be added to initial ideas and maintained into the later stages. As a result, the final design solution, which is analysed and optimised, may be superior to the single solution in the TDE.

Characteristics of parametric design thinking

Woodbury (2010) defines parametric design thinking using three characteristics: thinking with abstraction, thinking mathematically and thinking algorithmically. Thinking with abstraction provides the foundation for generating numbers of alternatives and reusing parametric model parts. Thinking mathematically relies on coding theorems and constructions into scripting paragraphs. Thinking algorithmically means that the scripting language provides functions that can add, modify or erase objects in a parametric model. The basic goal of parametric thinking is to establish the data-flow route clearly and precisely.

Woodbury's characteristics of parametric design thinking are focused more on the mathematical aspects of modelling. They suggest that in a PDE designers need a different kind of geometric knowledge that can understand and be used to structure the mathematical toolbox that supports their design intention (Woodbury, 2010). Thus, designers need more than just architectural knowledge, they need specialist knowledge of the mathematical tools used in the design process. There should, however, be a balance between knowledge of tool capability and architectural knowledge in the parametric design process. The danger of a lack of balance can be seen in many works arising from *parametricism* (Schumacher, 2009), where designers display too much knowledge of tool capability and too little of basic architectural principles. Neither the production of formal novelty, nor the celebration of the power of the software associated with it, are a substitute for functional or responsive design (Ostwald, 2004; Ostwald, 2010; Ostwald, 2015). Parametric design tools are too often only used to generate forms, not to capture people's psychological needs, social and historical impact or environmental concerns (Castellano, 2011).

Aranda and Lasch (2008) suggest that parametric design communicates between two worlds. The first, entirely abstract and coded, from which complex spatial worlds could emerge through simple mathematical expressions governed by parameters. The second is the one we find through our everyday interactions with people, communities and cities (Aranda and Lasch, 2008). Aish (2005) proposes two levels of algorithmic thinking. In the first level, there is a desire to explore geometric subtleties in which equations are established to describe modelling relationships, while the other level captures a desire to control "unpredictability" over large data sets, wherein associative data sometimes emerges from previously unexplored conditions (Aish, 2005). Aish's first level of algorithmic thinking can be defined as geometrically based, manually controlled and predictable, while the second level can be defined as data-based, automatically generated and unpredictable. Generally speaking, maintaining a balance between algorithmic thinking and architectural or design thinking is very important in parametric design. But this does raise the question, that if architects are familiar with architectural design thinking, how should algorithmic thinking be developed and integrated with architectural thinking in a PDE? Thus far, no clear answer is available to this question, but some of the studies mentioned previously in this chapter and those presented in Chapter 4 do begin to consider the issue.

In summary, researchers have identified two important aspects of parametric design thinking. The first is that it is abstract and rule-oriented, the second is that it is dynamic, lively and design-oriented. When using parametric tools, designers need to maintain a fine balance between these two, not being overly dominated by codes and algorithms, but using the power of parametric design to aid architectural thinking.

Two levels of parametric design thinking

From the review of parametric design, we can see that PDEs are typically presented as being different from TDEs due to their reliance on rules and algorithms. In addition to documentation and modelling, rule algorithms are core to design activities in PDEs, assisting designers by generating design paradigms and constructing data structures (Iordanova et al., 2009). However, the ways in which parametric design is used by architects are not well understood, which is why some argue that parametric designers require a deeper knowledge of how parametric tools can support their design intentions (Sanguinetti and Kraus, 2011).Compared to TDEs, in PDEs architects not only design by applying specialist knowledge, but they also explicitly define rules and their logical relationships using parameters (Abdelsalam, 2009). When an architect models a building form using parameters they must assess variations, design data-flow routes, adjust the values of parameters and revise rules (if necessary). At this time, they are not only thinking about the particular building design but also about the rule design. It is through the control of logical relationships between forms and functions that the possibilities for design solutions are heightened (Hernandez, 2006; Karle and Kelly, 2011).

In a typical parametric design process, design activities occur on two levels: the design knowledge level and the rule algorithm level (Figure 3.1). At the design knowledge level, architects make use of their design knowledge, including for example, how to adapt a building to the site, how to shape the way people use a building, and how to satisfy the requirements of their clients. At the rule algorithm level, designers apply design knowledge through the operations of the parametric design tools, including defining the rules and their logical relationships, choosing parameters suitable for a particular purpose and importing external data into the proposed rules. During the design process, designers apply specialist knowledge by defining rules and their logical relationships, which is known as *parameterisation*. These two levels of thinking are essential for understanding design cognition in PDEs.

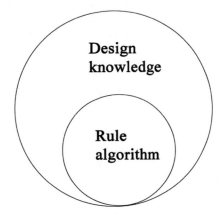

Figure 3.1. Two levels of parametric design thinking.

3.4. Conclusion

This chapter has reviewed the methods and examples of design cognition studies. It has examined design cognition from three perspectives: design thinking, design-problem and -solution spaces, and design creativity. Design creativity is a core issue in design studies. With emerging computational design environments, the ways in which design creativity is supported and expressed is critical. The second half of the chapter focused on design cognition in computational design environments, in particular, computational thinking and design cognition in PDEs. In generative and parametric design environments, designers not only use design knowledge, but also create rules to achieve their design intentions. This chapter also provided a foundation for the studies presented in

Chapter 4, which includes relevant theoretical underpinnings, formal approaches and methods, as well as benchmarking studies for better understanding of how designers think and behave in different computational design environments.

References

Abdelhameed, W. (2009). Cognition model in conceptual designing. pp. 771–780. *In*: *Proceedings of the 14th International Conference on Computer Aided Architectural Design Research*. Taiwan.

Abdelsalam, M. (2009). The use of the smart geometry through various design processes: Using the programming platform (parametric features) and generative components. *In*: *Proceedings of the Arab Society for Computer Aided Architectural Design (ASCAAD 2009)*, Manama, Kingdom of Bahrain. http://papers.cumincad.org/cgi-bin/works/paper/ascaad2009_mai_abdelsalam

Ahmed, S. (2007). Empirical research in engineering practice. *Journal of Design Research*, 6(3), 359–380. https://doi.org/10.1504/JDR.2007.016389

Aish, R. (2005). From intuition to precision. pp. 10–14. *In*: *Proceedings of 23rd eCAADe Conference*. Lisbon, Portugal.

Akin, O. (1986). *Psychology of Architectural Design*. Pion.

Akın, Ö. (2001). Variants of design cognition. pp. 105–124. *In*: C. Eastman, W. Newstetter and M. McCracken (Eds.), Design Knowing and Learning: Cognition in Design Education. Elsevier.

Akin, Ö. and Lin, C. (1995). Design protocol data and novel design decisions. *Design Studies*, 16(2), 211–236. https://doi.org/10.1016/0142-694x(94)00010-b

Akin, O. and Moustapha, H. (2004). Strategic use of representation in architectural massing. *Design Studies*, 25(1), 31–50. https://doi.org/https://doi.org/10.1016/S0142-694X(03)00034-6

Al-Sayed, K., Dalton, R.C. and Hölscher, C. (2010). Discursive design thinking: The role of explicit knowledge in creative architectural design reasoning. *Artificial Intelligence for Engineering Design, Analysis and Manufacturing: AIEDAM*, 24(2), 211–230. https://doi.org/10.1017/S0890060410000065

Aldous, C. (2005). Creativity in problem solving: Uncovering the origin of new ideas. *International Education Journal*, 5(ERC2004 Special Issue), 43–56.

Almendra, R.A. and Christiaans, H. (2009). Decision making in the conceptual design phases: A comparative study. *Journal of Design Research*, 8(1), 1–22. https://doi.org/10.1504/JDR.2009.030997

Amabile, T.M. (1982). Social psychology of creativity: A consensual assessment technique. *Journal of Personality and Social Psychology*, 43, 997–1013. https://doi.org/10.1037/0022-3514.43.5.997

Amabile, T.M. (1983). *The Social Psychology of Creativity*. Springer-Verlag.

Aranda, B. and Lasch, C. (2008). What is parametric to us. pp. 195. *In*: T. Sakamoto and A. Ferré (Eds.), From Control to Design: Parametric/Algorithmic Architecture. Actar-D.

Arnheim, R. (1974). *Art and Visual Perception: A Psychology of the Creative Eye*. University of California Press.

Asimow, M. (1962). *Introduction to Design.* Prentice-Hall.

Atman, C.J. (2019). Design timelines: Concrete and sticky representations of design process expertise. *Design Studies,* 65, 125–151. https://doi.org/10.1016/j.destud.2019.10.004

Atman, C.J. and Bursic, K.M. (1996). Teaching engineering design: Can reading a textbook make a difference? *Research in Engineering Design— Theory, Applications, and Concurrent Engineering,* 8(4), 240–250. https://doi.org/10.1007/BF01597230

Atman, C.J. and Bursic, K.M. (1998). Verbal protocol analysis as a method to document engineering student design processes. *Journal of Engineering Education,* 87(2), 121–132. https://doi.org/10.1002/j.2168-9830.1998.tb00332.x

Atman, C.J., Chimka, J.R., Bursic, K.M. and Nachtmann, H.L. (1999). A comparison of freshman and senior engineering design processes. *Design Studies,* 20(2), 131-152. https://doi.org/http://dx.doi.org/10.1016/S0142-694X(98)00031-3

Atman, C.J., Cardella, M.E., Turns, J. and Adams, R. (2005). Comparing freshman and senior engineering design processes: An in-depth follow-up study. *Design Studies,* 26(4), 325–357. https://doi.org/10.1016/j.destud.2004.09.005

Auffrey, C. and Hildebrandt, H. (2014). Utilizing 3M's Visual Attention Service software to assess on-premise signage conspicuity in complex signage environments found in urban neighborhood and suburban strip business districts: Lessons learned from a graduate seminar [Conference paper]. 2014 National Signage Research and Education Conference.

Azevedo, R. and Jacobson, M.J. (2008). Advances in scaffolding learning with hypertext and hypermedia: A summary and critical analysis. *Educational Technology Research and Development,* 56(1), 93–100. https://doi.org/10.1007/s11423-007-9064-3

Bell, P. (2005). Reflections on the Cognitive and Social Foundations of Information and Communication Technology Fluency. Paper read at Workshop on ICT Fluency and High School Graduation Outcomes. Washington, D.C.

Bernal, M., Haymaker, J.R. and Eastman, C. (2015). On the role of computational support for designers in action. *Design Studies,* 41, 163–182. https://doi.org/https://doi.org/10.1016/j.destud.2015.08.001

Bertoni, A. (2013). Analyzing product-service systems conceptual design: The effect of color-coded 3D representation. *Design Studies,* 34(6), 763–793. http://www.scopus.com/inward/record.url?eid=2-s2.0-84884594523&partnerID=40&md5=03827a1f733bbcc661002a309b50a54d

Besemer, S.P. and O'Quin, K. (1993). Assessing creative products: Progress and potentials. *In:* Isaksen, S.G., Murdock, M.C., Firestien, R.L. and Treffinger, D.J. (Eds.), Nurturing and Developing Creativity: The Emergence of a Discipline. Ablex.

Bilda, Z. and Demirkan, H. (2003). An insight on designers' sketching activities in traditional versus digital media. *Design Studies,* 24(1), 27–50. https://doi.org/10.1016/s0142-694x(02)00032-7

Bilda, Z. and Gero, J.S. (2007). The impact of working memory limitations on the design process during conceptualization. *Design Studies,* 28(4), 343–367. https://doi.org/10.1016/j.destud.2007.02.005

Bilda, Z., Gero, J.S. and Purcell, T. (2006). To sketch or not to sketch? That is the question. *Design Studies,* 27(5), 587–613. https://doi.org/10.1016/j.destud.2006.02.002

Blom, N. and Bogaers, A. (2020). Using Linkography to investigate students' thinking and information use during a STEM task. *International Journal of Technology and Design Education*, 30(1). https://doi.org/10.1007/s10798-018-9489-5

Boden, M.A. (2004). *The Creative Mind: Myths And Mechanisms* (2nd ed.). Routledge.

Boland, R.J., Collopy, F.L., Lyytinen, K. and Yoo, Y. (2008). Managing as designing: Lessons for organization leaders from the design practice of Frank O. Gehry. *Design Issues*, 24(1), 10–25.

Brösamle, M. and Hölscher, C. (2018). Approaching the architectural native: A graphical transcription method to capture sketching and gesture activity. *Design Studies*, 56, 1–27. https://doi.org/10.1016/j.destud.2018.01.002

Brown, D.C. (2012). Creativity, Surprise and Design: An Introduction and Investigation [Conference paper]. The 2nd International Conference on Design Creativity (ICDC2012), Glasgow, UK.

Brown, T. and Katz, B. (2009). *Change by Design: How Design Thinking Transforms Organizations and Inspires Innovation*. Harper Collins Publishers.

Bruner, J.S. (1962). The conditions of creativity. pp. 1–30. *In*: H. Gruber, G. Terrell and M. Wertheimer (Eds.), Contemporary Approaches to Cognition. Atherton.

Cao, J., Xiong, Y., Li, Y., Liu, L. and Wang, M. (2018). Differences between beginning and advanced design students in analogical reasoning during idea generation: Evidence from eye movements. *Cognition, Technology and Work*, 20, 1–16.

Cardella, M.E., Atman, C.J. and Adams, R.S. (2006). Mapping between design activities and external representations for engineering student designers. *Design Studies*, 27(1), 5–24. https://doi.org/10.1016/j.destud.2005.05.001

Casakin, H. (2019). Metaphors as discourse interaction devices in architectural design. *Buildings*, 9(2), Article 52. https://doi.org/10.3390/buildings9020052

Cash, P., Hicks, B. and Culley, S. (2015). Activity theory as a means for multi-scale analysis of the engineering design process: A protocol study of design in practice. *Design Studies*, 38, 1–32. https://doi.org/10.1016/j.destud.2015.02.001

Cash, P. and Kreye, M. (2018). Exploring uncertainty perception as a driver of design activity. *Design Studies*, 54, 50–79. https://doi.org/10.1016/j.destud.2017.10.004

Cash, P. and Maier, A. (2016). Prototyping with your hands: The many roles of gesture in the communication of design concepts. *Journal of Engineering Design*, 27(1–3), 118–145. https://doi.org/10.1080/09544828.2015.1126702

Castellano, D. (2011). Humanizing parametricism. *In*: J. Cheon, S.n. Hardy and T. Hemsath (Eds.), *Proceedings of ACADIA Regional 2011 Conference*. http://papers.cumincad.org/data/works/att/acadiaregional2011_032.content.pdf

Chai, K.–H. and Xiao, X. (2012). Understanding design research: A bibliometric analysis of design studies (1996–2010). *Design Studies*, 33(1), 24–43. https://doi.org/http://dx.doi.org/10.1016/j.destud.2011.06.004

Chakrabarti, A., Morgenstern, S. and Knaab, H. (2004). Identification and application of requirements and their impact on the design process: A protocol study. *Research in Engineering Design*, 15(1), 22–39. https://doi.org/10.1007/s00163-003-0033-5

Chakrabarti, A., Sarkar, P., Leelavathamma, B. and Nataraju, B.S. (2005). A functional representation for aiding biomimetic and artificial inspiration of new ideas. *Artificial Intelligence for Engineering Design, Analysis and Manufacturing*, 19(2), 113–132. https://doi.org/10.1017/S0890060405050109

Chan, C.-S. (2001). An examination of the forces that generate a style. *Design Studies*, 22(4), 319–346. https://doi.org/https://doi.org/10.1016/S0142-694X(00)00045-4

Chandrasekera, T., Vo, N. and D'Souza, N. (2013). The effect of subliminal suggestions on Sudden Moments of Inspiration (SMI) in the design process. *Design Studies*, 34(2), 193–215. http://www.scopus.com/inward/record.url?eid=2-s2.0-84873466714&partnerID=40&md5=addd4603e5830ce40f824365eca6383a

Chen, S.-C. (2001). The role of design creativity in computer media. *In*: *Proceedings of the 19th Education and Research in Computer Aided Architectural Design in Europe*, Helsinki, Finland. http://www.ecaade.org/prev-conf/archive/ecaade2001/site/E2001presentations/09_04_chen.pdf

Cheong, H., Hallihan, G.M. and Shu, L.H. (2014). Design problem solving with biological analogies: A verbal protocol study [Article]. *Artificial Intelligence for Engineering Design, Analysis and Manufacturing: AIEDAM*, 28(1), 27–47. https://doi.org/10.1017/S0890060413000486

Chien, S.-F. and Yeh, Y.-T. (2012). On creativity and parametric design—A preliminary study of designer's behaviour when employing parametric design tools. pp. 245–253. *In*: H. Achten, J. Pavlíček and J. Huhín (Eds.), *Proceedings of 30th International Conference on Education and Research in Computer Aided Architectural Design in Europe (eCAADe 2012)*. Czech Republic. http://papers.cumincad.org/cgi-bin/works/Show?ecaade2012_223

Chu, P.Y., Hung, H.Y., Wu, C.F. and Liu, Y.T. (2017). Effects of various sketching tools on visual thinking in idea development. *International Journal of Technology and Design Education*, 27(2), 291–306. https://doi.org/10.1007/s10798-015-9349-5

Coley, F., Houseman, O. and Roy, R. (2007). An introduction to capturing and understanding the cognitive behaviour of design engineers. *Journal of Engineering Design*, 18(4), 311–325. https://doi.org/10.1080/09544820600963412

Cramer-Petersen, C.L., Christensen, B.T. and Ahmed-Kristensen, S. (2019). Empirically analysing design reasoning patterns: Abductive-deductive reasoning patterns dominate design idea generation. *Design Studies*, 60, 39–70. https://doi.org/10.1016/j.destud.2018.10.001

Cropley, A.J. (1999). Definitions of creativity. pp. 511–524. *In*: M.A. Runco and S.R. Pritzker (Eds.), Encyclopedia of Creativity (Vol. 1), Academic Press. http://books.google.com/books?id=_dK79AdKmIoC

Cropley, D. and Cropley, A. (2005). Engineering creativity: A systems concept of functional creativity. pp. 169–185. *In*: J. Kaufman and J. Baer (Eds.), *Creativity Across Domains*. Lawrence Erlbaum Associates.

Cross, N. (2001). Design cognition: Results from protocol and other empirical studies of design activity. pp. 79–103. *In*: E. Charles, M. Michael and N. Wendy (Eds.), *Design Knowing and Learning: Cognition in Design Education*. Elsevier Science. https://doi.org/10.1016/b978-008043868 9/50005-x

Cross, N. (2011). *Design Thinking: Understanding How Designers Think and Work* (English ed.). Berg Publishers.

Cross, N. and Cross, C. (1998). Expertise in engineering design. *Research in Engineering Design*, 10, 141–149.

D'Souza, N. and Dastmalchi, M.R. (2016). Creativity on the move: Exploring little-c (p) and big-C (p) creative events within a multidisciplinary design team process. *Design Studies*, 46, 6–37. https://doi.org/10.1016/j.destud.2016.07.003

Davis, D.C., Gentili, K.L., Trevisan, M.S. and Calkins, D.E. (2002). Engineering design assessment processes and scoring scales for program improvement and accountability. *Journal of Engineering Education*, 91(2), 211–221. https://doi.org/10.1002/j.2168-9830.2002.tb00694.x

Deken, F., Kleinsmann, M., Aurisicchio, M., Lauche, K. and Bracewell, R. (2012). Tapping into past design experiences: Knowledge sharing and creation during novice-expert design consultations. *Research in Engineering Design*, 23(3), 203–218. https://doi.org/10.1007/s00163-011-0123-8

Dewett, T. (2003). Understanding the relationship between information technology and creativity in organizations. *Creativity Research Journal*, 15, 167–182.

Dinar, M., Shah, J.J., Cagan, J., Leifer, L., Linsey, J., Smith, S.M. and Hernandez, N.V. (2015). Empirical studies of designer thinking: Past, present, and future. *Journal of Mechanical Design*, 137(2). https://doi.org/10.1115/1.4029025

diSessa, A. (2005). Systemics of learning for a revised pedagogical agenda. *In*: R.L. Mahwah (Ed.), *Foundations for the Future in Mathematics Education*. Lawrence Erlbaum.

Dixon, R.A. and Bucknor, J. (2019). A comparison of the types of heuristics used by experts and novices in engineering design ideation. *Journal of Technology Education*, 30(2), 39–59. https://doi.org/10.21061/jte.v30i2.a.3

Do, E.Y.–L. and Gross, M.D. (2001). Thinking with diagrams in architectural design. *Artificial Intelligence Review*, 15(1), 135–149. https://doi.org/10.1023/A:1006661524497

Dodge, R. and Cline, T.S. (1901). The angle velocity of eye movements. *Psychological Review*, 8, 145–157.

Dorst, K. and Cross, N. (2001). Creativity in the design process: Co-evolution of problem-solution. *Design Studies*, 22(5), 425–437. https://doi.org/10.1016/s0142-694x(01)00009-6

Dorst, K. and Dijkhuis, J. (1995). Comparing paradigms for describing design activity. *Design Studies*, 16(2), 261–274. https://doi.org/10.1016/0142-694x(94)00012-3

Eastman, C.M. (1970). *On the Analysis of Intuitive Design Process*. The MIT Press.

Ekströmer, P. (2019). *A First Sketch of Computer Aided Ideation: Exploring CAD Tools as Externalization Media in Design Ideation*. https://doi.org/10.3384/lic.diva-162022

Ensici, A., Badke-Schaub, P., Bayazit, N. and Lauche, K. (2013). Used and rejected decisions in design teamwork. *CoDesign*, 9(2), 113–131. https://doi.org/10.1080/15710882.2013.782411

Ericsson, K.A. and Simon, H.A. (1993). *Protocol Analysis: Verbal Reports as Data*. MIT Press.

Eseryel, D., Ifenthaler, D. and Ge, X. (2013). Validation study of a method for assessing complex ill–structured problem solving by using causal representations. *Educational Technology Research and Development*, 61(3), 443–463. https://doi.org/10.1007/s11423-013-9297-2

Ferreira, J., Christiaans, H. and Almendra, R. (2016). A visual tool for analysing teacher and student interactions in a design studio setting. *CoDesign*, 12(1–2), 112–131. https://doi.org/10.1080/15710882.2015.1135246

Frankenberger, E. and Auer, P. (1997). Standardized observation of team-work in design. *Research in Engineering Design—Theory, Applications, and Concurrent Engineering*, 9(1), 1–9. https://doi.org/10.1007/BF01607053

Galle, P. and Béla Kovács, L. (1996). Replication protocol analysis: A method for the study of real-world design thinking. *Design Studies*, 17(2), 181–200. https://doi.org/https://doi.org/10.1016/0142-694X(95)00039-T

Gero, J. (2000). Computational models of innovative and creative design processes. *Technological Forecasting and Social Change*, 64(2–3), 183–196. https://doi.org/10.1016/s0040-1625(99)00105-5

Gero, J. and Tang, H.-H. (2001). The differences between retrospective and concurrent protocols in revealing the process-oriented aspects of the design process. *Design Studies*, 22(3), 283–295. https://doi.org/10.1016/s0142-694x(00)00030-2

Gero, J.S. (1990). Design prototypes: A knowledge representation schema for design. *AI Magazine*, 11(4), 26–36.

Gero, J.S. (1996). Creativity, emergence and evolution in design. *Knowledge-Based Systems*, 9(7), 435–448. https://doi.org/10.1016/s0950-7051(96)01054-4

Gero, J.S. and McNeill, T. (1998). An approach to the analysis of design protocols. *Design Studies*, 19(1), 21–61. https://doi.org/10.1016/s0142-694x(97)00015-x

Gero, J.S. and Tang, H.-H. (1999). Concurrent and retrospective protocols and computer-aided architectural design. pp. 403–410. *In*: Z. Xie and J. Quian (Eds.), Fourth Conference on Computer Aided Architectural Design Research in Asia (CAADRIA1999). Shanghai.

Gero, J.S. and Kan, J. (2016). Empirical results from measuring design creativity: Use of an augmented coding scheme in protocol analysis. *In*: J. Linsey, M. Yang and Y. Nagai (Eds.), DS86: Proceedings of the 4th International Conference on Design Creativity. Georgia Institute of Technology, Atlanta, Georgia.

Goldschmidt, G. (1989). Problem representation versus domain of solution in architectural design teaching. *Architectural and Planning Research*, 6, 204–215.

Goldschmidt, G. (1991). The dialectics of sketching. *Creativity Research Journal*, 4(2), 123–143.

Goldschmidt, G. (2016). Linkographic evidence for concurrent divergent and convergent thinking in creative design. *Creativity Research Journal*, 28(2), 115–122. https://doi.org/10.1080/10400419.2016.1162497

Goldschmidt, G., Hochman, H. and Dafni, I. (2010). The design studio crit: Teacher-student communication. *Artificial Intelligence for Engineering Design, Analysis and Manufacturing: AIEDAM*, 24(3), 285–302. https://doi.org/10.1017/S089006041000020X

Gonzalez-Sanchez, J., Baydogan, M., Chavez-Echeagaray, M.E., Atkinson, R.K. and Burleson, W. (2017). Affect measurement: A roadmap through approaches, technologies, and data analysis. pp. 255–288. *In*: M. Jeon (Ed.), Emotions and Affect in Human Factors and Human-computer Interaction. Academic Press. https://doi.org/https://doi.org/10.1016/B978-0-12-801851-4.00011-2

Gould, D. and Peeples, R. (1970). Eye movements during visual search and discrimination of meaningless symbol and object patterns. *Journal of Experimental Psychology*, 85, 51–55.

Grubbs, M.E., Strimel, G.J. and Kim, E. (2018). Examining design cognition coding schemes for P-12 engineering/technology education. *International Journal of Technology and Design Education*, 28(4), 899–920. https://doi.org/10.1007/s10798-017-9427-y

Guan, Z., Lee, S., Cuddihy, E. and Raney, J. (2006). The validity of the stimulated retrospective think-aloud method as measured by eye tracking. pp. 1253–1262. *In*: *Proceedings of the CHI 2006 Conference on Human Factors in Computing Systems*. New York. https://doi.org/10.1145/1124772.1124961

Guidera, S. and MacPherson, D.S. (2008). Digital modeling in design foundation coursework: An exploratory study of the effectiveness of conceptual design software. *The Journal of Technology Studies*, 34(1), 55–66. www.jstor.org/stable/43604226

Guilford, J.P. (1975). Creativity: A quarter century of progress. pp. 37–59. *In*: A. Taylor and J.W. Getzels (Eds.), Perspectives in Creativity. Aldine.

Guilford, J.P. (1967). *The Nature of Human Intelligence*. McGraw-Hill.

Hasirci, D. and Demirkan, H. (2007). Understanding the effects of cognition in creative decision making: A creativity model for enhancing the design studio process. *Creativity Research Journal*, 19(2–3), 259–271. https://doi.org/10.1080/10400410701397362

Hay, L., Duffy, A.H.B., McTeague, C., Pidgeon, L.M., Vuletic, T. and Grealy, M. (2017a). A systematic review of protocol studies on conceptual design cognition: Design as search and exploration. *Design Science*, 3. https://doi.org/10.1017/dsj.2017.11

Hay, L., Duffy, A.H.B., McTeague, C., Pidgeon, L.M., Vuletic, T. and Grealy, M. (2017b). Towards a shared ontology: A generic classification of cognitive processes in conceptual design. *Design Science*, 3. https://doi.org/10.1017/dsj.2017.6

Hayes, J.R. (1978). *Cognitive Psychology: Thinking and Creating*. Dorsey Press. http://books.google.com/books?id=l1N-AAAAMAAJ

Hernandez, C.R.B. (2006). Thinking parametric design: Introducing parametric Gaudi. *Design Studies*, 27(3), 309–324. https://doi.org/10.1016/j.destud.2005.11.006

Ho, C.-H. (2001). Some phenomena of problem decomposition strategy for design thinking: Differences between novices and experts. *Design Studies*, 22(1), 27–45. https://doi.org/https://doi.org/10.1016/S0142-694X(99)00030-7

Hocevar, D. (1981). Measurement of creativity: Review and critique. *Journal of Personality Assessment*, 45, 450–464.

Holland, N. (2011). Inform form perform. pp. 131–140. *In*: J. Cheon, S.n. Hardy and T. Hemsath (Eds.), *Proceedings of ACADIA Regional 2011 Conference*. https://digitalcommons.unl.edu/arch_facultyschol/21

Horgen, T., Joroff, M.L., Schon, D.A. and Porter, W.L. (1999). *Excellence by Design: Transforming Workplace and Work Practice*. John Wiley and Sons.

Houseman, O., Coley, F. and Roy, R. (2008). Comparing the cognitive actions of design engineers and cost estimators. *Journal of Engineering Design*, 19(2), 145–158. https://doi.org/10.1080/09544820701802964

Hu, Y., Du, X., Bryan-Kinns, N. and Guo, Y. (2019). Identifying divergent design thinking through the observable behavior of service design novices. *International Journal of Technology and Design Education*, 29(5), 1179–1191. https://doi.org/10.1007/s10798-018-9479-7

Huang, W., Matsushita, D. and Munemoto, J. (2012). Protocol analysis of designers using an interactive evolutionary computation. *Frontiers of Architectural Research*, 1(1), 44–50. https://doi.org/10.1016/j.foar.2012.02.003

Ibrahim, R. and Pour Rahimian, F. (2010). Comparison of CAD and manual sketching tools for teaching architectural design. *Automation in Construction*, 19(8), 978–987. https://doi.org/10.1016/j.autcon.2010.09.003

Iordanova, I., Tidafi, T., Guité, M., De Paoli, G. and Lachapelle, J. (2009). Parametric methods of exploration and creativity during architectural design: A case study in the design studio. pp. 423–439. *In: Proceedings of the 13th International Conference on Computer Aided Architectural Design Futures*. Montréal.

Jacob, R. and Karn, K. (2003). Eye tracking in human–computer interaction and usability research: Ready to deliver the promises. pp. 573–605. In: J. Hyönä, R. Radach and H. Deubel (Eds.), The Mind's Eye: Cognitive and Applied Aspects of Eye Movement Research. Elsevier Science BV.

Jagtap, S. (2018). Design creativity: Refined method for novelty assessment. *International Journal of Design Creativity and Innovation*, 1–17. https://doi.org/1 0.1080/21650349.2018.1463176

Jagtap, S., Larsson, A., Hiort, V., Olander, E., Warell, A. and Khadilkar, P. (2014). How design process for the base of the pyramid differs from that for the top of the pyramid. *Design Studies*, 35(5), 527–558. https://doi.org/10.1016/j.destud.2014.02.007

Jiang, H. (2012). Understanding Senior Design Students' Product Conceptual Design Activities—A Comparison between Industrial and Engineering Design Students [Doctoral dissertation]. National University of Singapore.

Jiang, H., Gero, J. and Yen, C.C. (2014). Exploring designing styles using a problem-solution index. *In*: J.S. Gero (Ed.), *Proceedings of the Fifth International Conference of Design Computing and Cognition (DCC'12)*, College Station, USA.

Jin, Y. and Benami, O. (2010). Creative patterns and stimulation in conceptual design. *Artificial Intelligence for Engineering Design, Analysis and Manufacturing: AIEDAM*, 24(2), 191–209. https://doi.org/10.1017/S0890060410000053

Jin, Y. and Chusilp, P. (2006). Study of mental iteration in different design situations. *Design Studies*, 27(1), 25–55. https://doi.org/10.1016/j.destud.2005.06.003

Jordanous, A. (2011). Evaluating evaluation: Assessing progress in computational creativity research. *In: Proceedings of the Second International Conference on Computational Creativity (ICCC-11)*, Mexico City, Mexico.

Kafura, D. and Tatar, D. (2011). Initial Experience with a Computational Thinking Course for Computer Science Students [Conference paper]. SIGCSE '11, March 9–12, Dallas, Texas.

Kan, J.W.T. and Gero, J.S. (2008). Acquiring information from linkography in protocol studies of designing. *Design Studies*, 29(4), 315–337. https://doi.org/10.1016/j.destud.2008.03.001

Kan, J.W.T. and Gero, J.S. (2009). Using the FBS ontology to capture semantic design information in design protocol studies. pp. 213–229. *In*: J. McDonnell and P. Lloyd (Eds.), About: Designing. Analysing Design Meetings. Taylor and Francis.

Kan, J.W.T. and Gero, J.S. (2010). Exploring quantitative methods to study design behavior in collaborative virtual workspaces. pp. 273–282. *In*: B. Dave, A. Li, N. Gu and H-J. Park (Eds.), *Proceedings of CAADRIA 2010*.

Kan, J.W.T. and Gero, J.S. (2018). Characterizing innovative processes in design spaces through measuring the information entropy of empirical data from protocol studies. *Artificial Intelligence for Engineering Design, Analysis and Manufacturing: AIEDAM*, 32(1), 32–43. https://doi.org/10.1017/S0890060416000548

Kan, J.W.T., Gero, J.S. and Tang, H.-H. (2011). Measuring cognitive design activity changes during an industry team brainstorming session. pp. 621–640. *In*: J.S. Gero (Ed.), *Design Computing and Cognition '10*. Springer Netherlands. https://doi.org/10.1007/978-94-007-0510-4_33

Kannengiesser, U. and Gero, J.S. (2017). Can Pahl and Beitz' systematic approach be a predictive model of designing? *Design Science*, 3, Article e24. https://doi.org/10.1017/dsj.2017.24

Karle, D. and Kelly, B. (2011). Parametric thinking. pp. 109–113. *In*: J. Cheon, S.n. Hardy and T. Hemsath (Eds.), *Proceedings of ACADIA Regional 2011 Conference*.

Kaufman, J.C. and Sternberg, R.J. (2006). *The International Handbook of Creativity*. Cambridge University Press.

Kaufman, L. and Richard, W. (1969). Spontaneous fixation tendencies for visual forms. *Perception and Psychophysics*, 5, 85–88.

Kavakli, M. (2001). NoDes: kNOwledge-based modeling for detailed DESign process—From analysis to implementation. *Automation in Construction*, 10(4), 399–416. https://doi.org/10.1016/S0926-5805(00)00076-5

Kavakli, M. and Gero, J.S. (2002). The structure of concurrent cognitive actions: A case study on novice and expert designers. *Design Studies*, 23(1), 25–40. https://doi.org/10.1016/s0142-694x(01)00021-7

Kelley, T. and Sung, E. (2017). Examining elementary school students' transfer of learning through engineering design using think-aloud protocol analysis. *Journal of Technology Education*, 28(2), 83–108. https://doi.org/10.21061/jte.v28i2.a.5

Kelley, T.R., Capobianco, B.M. and Kaluf, K.J. (2015). Concurrent think-aloud protocols to assess elementary design students. *International Journal of Technology and Design Education*, 25(4), 521–540. https://doi.org/10.1007/s10798-014-9291-y

Kim, E.J. and Kim, K.M. (2015). Cognitive styles in design problem solving: Insights from network-based cognitive maps. *Design Studies*, 40, 1–38. https://doi.org/10.1016/j.destud.2015.05.002

Kim, M.H., Kim, Y.S., Lee, H.S. and Park, J.A. (2007). An underlying cognitive aspect of design creativity: Limited commitment mode control strategy. *Design Studies*, 28(6), 585–604. https://doi.org/10.1016/j.destud.2007.04.006

Kim, M.J. (2006). The Effects of Tangible User Interfaces on Designers' Spatial Cognition [Doctoral dissertation]. University of Sydney.

Kim, M.J. and Maher, M.L. (2005). Creative design and spatial cognition in a tangible user interface environment. pp. 233–250. *In*: J. Gero and M.L. Maher (Eds.), *Computational and Cognitive Models of Creative Design VI*. University of Sydney.

Kim, M.J. and Maher, M.L. (2008). The impact of tangible user interfaces on spatial cognition during collaborative design. *Design Studies*, 29(3), 222–253. https://doi.org/10.1016/j.destud.2007.12.006

Kim, Y.S., Jin, S.T. and Lee, S.W. (2011). Relations between design activities and personal creativity modes. *Journal of Engineering Design*, 22(4), 235–257. https://doi.org/10.1080/09544820903272867

Kristensen, T. (2004). The physical context of creativity. *Creativity and Innovation Management*, 13, 89–96.

Kumar, J.S. and Bhuvaneswari, P. (2012). Analysis of electroencephalography (EEG) signals and its categorization—A study. *Procedia Engineering*, 38, 2525–2536. https://doi.org/https://doi.org/10.1016/j.proeng.2012.06.298

Kurtoglu, T., Swantner, A. and Campbell, M.I. (2008). Automating the conceptual design process: From black-box to component selection. *In*: J.S. Gero and A.K. Goel (Eds.), Design Computing and Cognition '08. Dordrecht.

Kuusela, H. and Pallab, P. (2000). A comparison of concurrent and retrospective verbal protocol analysis. *American Journal of Psychology*, 113(3), 387–404.

Lawson, B. (1997). *How Designers Think: The Design Process Demystified* (Revised 3rd ed.). Architectural Press.

Lawson, B. (2005). *How Designers Think* (4th ed.). Routledge.

Leblebici-Başar, D. and Altarriba, J. (2013). The role of imagery and emotion in the translation of concepts into product form. *Design Journal*, 16(3), 295–314. https://doi.org/10.2752/175630613X13660502571787

Lee, J., Ahn, J., Kim, J., Kho, J.M. and Paik, H.Y. (2018). Cognitive evaluation for conceptual design: Cognitive role of a 3D sculpture tool in the design thinking process. *Digital Creativity*, 29(4), 299–314. https://doi.org/10.1080/14626268.2018.1528988

Lee, J., Gu, N. and Williams, A. (2014). Parametric design strategies for the generation of creative designs. *International Journal of Architectural Computing*, 12(3), 263 282. https://doi.org/10.1260/1478-0771.12.3.263

Lee, J.H., Gu, N. and Ostwald, M.J. (2019). Cognitive and linguistic differences in architectural design. *Architectural Science Review*, 62(3), 248–260. https://doi.org/10.1080/00038628.2019.1606777

Lee, J.H., Ostwald, M.J. and Gu, N. (2016). The language of design: Spatial cognition and spatial language in parametric design. *International Journal of Architectural Computing*, 14(3), 277–288. https://doi.org/10.1177/1478077116663350

Lemons, G., Carberry, A., Swan, C., Jarvin, L. and Rogers, C. (2010). The benefits of model building in teaching engineering design. *Design Studies*, 31(3), 288–309. https://doi.org/10.1016/j.destud.2010.02.001

Liedtka, J., King, A. and Bennett, K. (2013). *Solving Problems with Design Thinking*. Columbia University Press.

Liikkanen, L.A. (2010). *Design Cognition for Conceptual Design*. Koneenrakennustekniikan laitos.

Liikkanen, L.A. and Perttula, M. (2009). Exploring problem decomposition in conceptual design among novice designers. *Design Studies*, 30(1), 38–59. https://doi.org/http://dx.doi.org/10.1016/j.destud.2008.07.003

Liu, L., Li, Y., Xiong, Y., Cao, J. and Yuan, P. (2018). An EEG study of the relationship between design problem statements and cognitive behaviors during conceptual design. *Artificial Intelligence for Engineering Design, Analysis and Manufacturing*, 32(3), 351–362. https://doi.org/10.1017/S0890060417000683

Liu, L., Nguyen, T.A., Zeng, Y. and Hamza, A.B. (2016). Identification of Relationships between Electroencephalography (EEG) Bands and Design Activities [Conference paper]. ASME 2016 International Design Engineering

Technical Conferences and Computers and Information in Engineering Conference.

Lohmeyer, Q., Matthiesen, S., Mussgnug, M. and Meboldt, M. (2014). Analysing visual behaviour in engineering design by eye tracking experiments. *In*: I. Horváth (Ed.), Tools and Methods of Competitive Engineering: Digital Proceedings of the Tenth International Symposium on Tools and Methods of Competitive Engineering – TMCE 2014, May 19–23, Budapest, Hungary.

López-Mesa, B., Mulet, E., Vidal, R. and Thompson, G. (2011). Effects of additional stimuli on idea-finding in design teams. *Journal of Engineering Design*, 22(1), 31–54. https://doi.org/10.1080/09544820902911366

Maher, M.L. (2010). Design creativity research: From the individual to the crowd. pp. 41–47. *In*: T. Taura and Y. Nagai (Eds.), Design Creativity 2010. https://link.springer.com/book/10.1007/978-0-85729-224-7

Maher, M.L. and Poon, J. (1996). Modelling design exploration as co-evolution. *Microcomputers in Civil Engineering*, 11(3), 195–210.

Maher, M.L., Poon, J. and Boulanger, S. (1996). Formalising design exploration as co-evolution. pp. 3–30. *In*: J.S. Gero and F. Sudweeks (Eds.), Advances in Formal Design Methods for CAD. *Proceedings of the IFIP WG5.2 Workshop on Formal Design Methods for Computer-Aided Design*, June 1995. Springer US. https://doi.org/10.1007/978-0-387-34925-1_1

Maher, M.L. and Tang, H.H. (2003). Co-evolution as a computational and cognitive model of design. *Research in Engineering Design*, 11, 47–63.

Mao, X., Galil, O., Parrish, Q. and Sen, C. (2020). Evidence of cognitive chunking in freehand sketching during design ideation. *Design Studies*, 67, 1–26. https://doi.org/10.1016/j.destud.2019.11.009

Martin, R. (2009). *The Design of Business: Why Design Thinking Is the Next Competitive Advantage*. Harvard Business Review Press.

Mc Neill, T., Gero, J.S. and Warren, J. (1998). Understanding conceptual electronic design using protocol analysis. *Research in Engineering Design— Theory, Applications, and Concurrent Engineering*, 10(3), 129–140. https://doi.org/10.1007/BF01607155

McKim, R.H. (1980). *Experiences in Visual Thinking* (2nd ed.). PWS Engineering.

Menezes, A. and Lawson, B. (2006). How designers perceive sketches. *Design Studies*, 27(5), 571–585. https://doi.org/https://doi.org/10.1016/j.destud.2006.02.001

Meniru, K., Rivard, H. and Bédard, C. (2003). Specifications for computer-aided conceptual building design. *Design Studies*, 24, 51–71. https://doi.org/10.1016/S0142-694X(02)00009-1

Mentzer, N., Becker, K. and Sutton, M. (2015). Engineering design thinking: High school students' performance and knowledge. *Journal of Engineering Education*, 104(4), 417–432. https://doi.org/10.1002/jee.20105

Mitchell, W.J. (Ed.). (2003). *Beyond Productivity: Information Technology, Innovation, and Creativity*. National Academies Press.

Mohamed Khaidzir, K.A. and Lawson, B. (2013). The cognitive construct of design conversation [Article]. *Research in Engineering Design*, 24(4), 331–347. https://doi.org/10.1007/s00163-012-0147-8

Moote, I. (2013). *Design Thinking for Strategic Innovation: What They Can't Teach You at Business or Design School*. Hoboken, NJ.

Mothersill, P. and Bove, V.M. Jr. (2018). An ontology of computational tools for design activities. pp. 1261–1277. *In*: C. Storni, K. Leahy, M. McMahon, E.

Bohemia and P. Lloyd (Eds.), *Proceedings of DRS 2018 International Conference*. Design Research Society.

Mothersill, P. and Bove, V.M. (2019). Beyond average tools. On the use of 'dumb' computation and purposeful ambiguity to enhance the creative process. *The Design Journal*, 22(sup1), 1147–1161. https://doi.org/10.1080/14606925.2019.1594981

Moursund, D. (2006). *Computational Thinking and Math Maturity: Improving Math Education in K–8 Schools* [Online document]. University of Oregon. https://pdfs.semanticscholar.org/b5a3/7c253eca73c0c99d095dbe9d4de5a6c1a681.pdf

Nguyen, L. and Shanks, G. (2009). A framework for understanding creativity in requirements engineering. *Information and Software Technology*, 51(3), 655–662. https://doi.org/10.1016/j.infsof.2008.09.002

Nguyen, P. (2017). *Approaches to Quantifying EEG Features for Design Protocol Analysis* [Doctoral dissertation]. Concordia University.

Nguyen, P., Nguyen, T.A. and Zeng, Y. (2015). Physiologically based segmentation of design protocol. *In: Proceedings of the 20th International Conference on Engineering Design (ICED 15)*, Milan, Italy.

Nguyen, P., Nguyen, T.A. and Zeng, Y. (2018). Empirical approaches to quantifying effort, fatigue and concentration in the conceptual design process. *Research in Engineering Design*, 29(3), 393–409. https://doi.org/10.1007/s00163-017-0273-4

Nguyen, P., Nguyen, T.A. and Zeng, Y. (2019). Segmentation of design protocol using EEG. *Artificial Intelligence for Engineering Design, Analysis and Manufacturing: AIEDAM*, 33(1), 11–23. https://doi.org/10.1017/S0890060417000622

Nguyen, T.A. and Zeng, Y. (2010). Analysis of design activities using EEG signals. *In: Proceedings of the ASME 2010 International Design Engineering Technical Conferences and Computers and Information in Engineering Conference*, Montreal, Canada.

Nguyen, T.A. and Zeng, Y. (2014). A physiological study of relationship between designer's mental effort and mental stress during conceptual design. *Computer-Aided Design*, 54, 3–18. https://doi.org/https://doi.org/10.1016/j.cad.2013.10.002

Nikander, J.B., Liikkanen, L.A. and Laakso, M. (2014). The preference effect in design concept evaluation. *Design Studies*, 35(5), 473–499. https://doi.org/10.1016/j.destud.2014.02.006

Oman, S. and Tumer, I. (2009). The potential of creativity metrics for mechanical engineering concept design. pp. 145–156. *In*: N. Bergendahl, M. Grimheden, L. Leifer, P. Skogstad and U. Lindemann (Eds.), *Proceedings of ICED 09, the 17th International Conference on Engineering Design (ICED'09)*, Palo Alto, CA, USA.

Önal, G.K. and Turgut, H. (2017). Cultural schema and design activity in an architectural design studio. *Frontiers of Architectural Research*, 6(2), 183–203. https://doi.org/10.1016/j.foar.2017.02.006

Osborn, A.F. (1963). *Applied Imagination: Principles and Procedures of Creative Problem-solving*. Scribner.

Ostwald, M. (2004). Freedom of form: Ethics and aesthetics in digital architecture. *The Philosophical Forum: Special Issue on Ethics and Architecture*, XXXV(2), 201–220.

Ostwald, M.J. (2010). Ethics and the auto-generative design process. *Building Research and Information: International Research, Development, Demonstration and Innovation*, 38(4), 390–400.

Ostwald, M.J. (2015). Ethics and geometry: Computational transformations and the curved surface in architecture. *In:* K. Williams and M.J. Ostwald (Eds.), Architecture and Mathematics: From Antiquity to the Future. Volume II: 1500s to the Future. Birkhäuser/Springer.

Park, J., Jin, Y., Ahn, S. and Lee, S. (2019). The impact of design representation on visual perception: Comparing eye-tracking data of architectural scenes between photography and line drawing. *Archives of Design Research*, 32(1), 5–29. https://doi.org/10.15187/ADR.2019.02.32.1.5

Parnes, S.J. (1981). *The Magic of Your Mind*. Bearly Limited.

Rahimian, F.P. and Ibrahim, R. (2011). Impacts of VR 3D sketching on novice designers' spatial cognition in collaborative conceptual architectural design. *Design Studies*, 32(3), 255–291. https://doi.org/10.1016/j.destud.2010.10.003

Rastogi, D. and Sharma, N.K. (2010). Creativity under concurrent and sequential task conditions. *Creativity Research Journal*, 22, 139–150.

Resnick, M., Maloney, J., Monroy Hernández, A., Rusk, N., Eastmond, E., Brennan, K., Millner, A., Rosenbaum, E.S.J., Silverman, B. and Kafai, Y. (2009). Scratch: Programming for all. *Communications of the ACM*, 52(11), 60–67.

Ritchie, G. (2007). Some empirical criteria for attributing creativity to a computer program. *Minds and Machines*, 17(1), 67–99.

Rosenman, M.A. and Gero, J.S. 1993. Creativity in design using a design prototype approach. *In:* J.S. Gero and L.M. Maher (Eds.), Modeling Creativity and Knowledge-based Creative Design. Hillsdale, New Jersey: Lawrence Erlbaum Associates.

Rowe, P.G. (1991). *Design Thinking*. MIT Press. http://books.google.com.au/books?id=ZjZ3mflzJtUC

Runco, M.A. (2004). Creativity. *Annual Review of Psychology*, 55, 657–687.

Runco, M.A. and Pritzker, S.R. (1999). *Encyclopedia of Creativity*. Academic Press.

Said-Metwaly, S., Noortgate, W.V.d. and Kyndt, E. (2017). Methodological issues in measuring creativity: A systematic literature review. *Creativity. Theories – Research - Applications*, 4(2), 276. https://doi.org/https://doi.org/10.1515/ctra-2017-0014

Sanguinetti, P. and Kraus, C. (2011). Thinking in parametric phenomenology. pp. 39–48. *In*: J. Cheon, S.n. Hardy and T. Hemsath (Eds.), *Proceedings of Parametricism: ACADIA Regional 2011 Conference*. Lincoln, Nebraska. https://digitalcommons.unl.edu/arch_facultyschol/21/

Sarkar, P. and Chakrabarti, A. (2008). Studying engineering design creativity – Developing a common definition and associated measures. *In: Proceedings of the NSF International Workshop on Studying Design Creativity'08—Design Science, Computer Science, Cognitive Science and Neuroscience Approaches: The State-of-the-Art*. University of Provence, Aix-en-Provence. https://cpdm.iisc.ac.in/cpdm/ideaslab/paper_scans/UID_17.pdf

Sarkar, P. and Chakrabarti, A. (2011). Assessing design creativity. *Design Studies*, 32(4), 348–383. https://doi.org/https://doi.org/10.1016/j.destud.2011.01.002

Sauder, J. and Jin, Y. (2016). A qualitative study of collaborative stimulation in group design thinking. *Design Science*, 2, Article e4. https://doi.org/10.1017/dsj.2016.1

Schön, D. and Wiggins, G. (1992). Kinds of seeing and their functions in designing. *Design Studies*, 13(2), 135–156. https://doi.org/10.1016/0142-694x(92)90268-f

Schön, D.A. (1983). *The Reflective Practitioner: How Professionals Think in Action*. Basic Books. http://books.google.com/books?id=ceJIWay4-jgC

Schumacher, P. (2009). Parametricism—A new global style for architecture and urban design. *AD Architectural Design—Digital Cities*, 79(4), 14–23.

Seitamaa-Hakkarainen, P. and Hakkarainen, K. (2001). Composition and construction in experts' and novices' weaving design. *Design Studies*, 22(1), 47–66. https://doi.org/10.1016/S0142-694X(99)00038-1

Self, J., Lee, S.-g. and Bang, H. (2014). Understanding the complexities of design representation. *In: Proceedings of 2013 Ancient Futures: Design and/or Happiness*. Asian Digital Art and Design Association and Korean Society of Design Science, Korea.

Shih, Y.T., Sher, W.D. and Taylor, M. (2017). Using suitable design media appropriately: Understanding how designers interact with sketching and CAD modelling in design processes. *Design Studies*, 53, 47–77. https://doi.org/10.1016/j.destud.2017.06.005

Sim, S.K. and Duffy, A.H.B. (2003). Towards an ontology of generic engineering design activities. *Research in Engineering Design*, 14(4), 200–223. https://doi.org/10.1007/s00163-003-0037-1

Sosa, R. and Gero, J.S. (2004). A computational framework for the study of creativity and innovation in design: Effects of social ties. *In*: J.S. Gero (Ed.), Design Computing and Cognition '04. Dordrecht.

Stempfle, J. and Badke-Schaub, P. (2002). Thinking in design teams—An analysis of team communication. *Design Studies*, 23(5), 473–496. https://doi.org/http://dx.doi.org/10.1016/S0142-694X(02)00004-2

Stonedahl, F., Wilkerson-Jerde, M. and Wilensky, U. (2009). Re-conceiving introductory computer science curricula through agent-based modeling. pp. 63–70. *In*: M. Beer, M. Fasli and D. Richards (Eds.), *Proceedings of the AAMAS 2009 Workshop on Educational Uses of Multi-Agent Systems (EduMas '09)*. https://www.cse.huji.ac.il/~jeff/aamas09/pdf/04_Workshop/w17.pdf

Stones, C. and Cassidy, T. (2007). Comparing synthesis strategies of novice graphic designers using digital and traditional design tools. *Design Studies*, 28(1), 59–72. https://doi.org/https://doi.org/10.1016/j.destud.2006.09.001

Strickfaden, M. and Heylighen, A. (2010). Scrutinizing design educators' perceptions of the design process. *Artificial Intelligence for Engineering Design, Analysis and Manufacturing: AIEDAM*, 24(3), 357–366. https://doi.org/10.1017/S0890060410000247

Sung, E. and Kelley, T.R. (2019). Identifying design process patterns: A sequential analysis study of design thinking. *International Journal of Technology and Design Education*, 29(2), 283–302. https://doi.org/10.1007/s10798-018-9448-1

Suwa, M., Gero, J. and Purcell, T. (1999). Unexpected discoveries and s-invention of design requirements: A key to creative designs. pp. 297–320. *In*: J.S. Gero and L.M. Maher (Eds.), Computational Models of Creative Design IV. Key Centre of Design Computing and Cognition.

Suwa, M., Purcell, T. and Gero, J. (1998). Macroscopic analysis of design processes based on a scheme for coding designers' cognitive actions. *Design Studies*, 19(4), 455–483. https://doi.org/10.1016/s0142-694x(98)00016-7

Suwa, M. and Tversky, B. (1996). What architects see in their sketches: Implications for design tools. *In*: Conference Companion on Human Factors in Computing Systems (CHI '96), New York, NY, USA.

Suwa, M. and Tversky, B. (1997). What do architects and students perceive in their design sketches? A protocol analysis. *Design Studies*, 18(4), 385–403. https://doi.org/10.1016/s0142-694x(97)00008-2

Tang, H.-H. and Gero, J.S. (2000). A content-oriented coding scheme for protocol analysis and computer-aided architectural design. *In*: *Proceedings of the Fifth Conference on Computer Aided Architectural Design Research in Asia (CAADRIA 2000)*, Singapore.

Tang, H.-H. and Gero, J.S. (2001). Sketches as affordances of meanings in the design process. *In*: J.S. Gero, B. Tversky and T. Purcell (Eds.), Visual and Spatial Reasoning in Design. Key Centre of Design Computing and Cognition, University of Sydney.

Tang, H.-H., Lee, Y.-Y. and Chiou, W.-K. (2009). Is an on-virtu digital sketching environment cognitively identical to in-sity free-hand sketching? *In*: *Proceedings of the 14th International Conference on Computer Aided Architectural Design Research in Asia*, Yunlin, Taiwan.

Tang, H.H., Lee, Y.Y. and Gero, J.S. (2011). Comparing collaborative co-located and distributed design processes in digital and traditional sketching environments: A protocol study using the function-behaviour-structure coding scheme. *Design Studies*, 32(1), 1–29. https://doi.org/10.1016/j.destud.2010.06.004

Taura, T., Yoshimi, T. and Ikai, T. (2002). Study of gazing points in design situation. *Design Studies*, 23, 165–185. https://doi.org/10.1016/S0142-694X(01)00018-7

Tedjosaputro, M.A., Shih, Y.T., Niblock, C. and Pradel, P. (2018). Interplay of sketches and mental imagery in the design ideation stage of novice designers. *Design Journal*, 21(1), 59–83. https://doi.org/10.1080/14606925.2018.1395655

Toh, C.A. and Miller, S.R. (2015). How engineering teams select design concepts: A view through the lens of creativity. *Design Studies*, 38, 111–138. https://doi.org/https://doi.org/10.1016/j.destud.2015.03.001

Torralba, A., Oliva, A., Castelhano, M.S. and Henderson, J.M. (2006). Contextual guidance of eye movements and attention in real-world scenes: The role of global features in object search. *Psychological Review*, 113(4), 766–786.

Torrance, E. (Ed.). (1966). *Torrance Test of Creative Thinking*. Princeton.

Tracey, M.W. and Hutchinson, A. (2019). Empathic design: Imagining the cognitive and emotional learner experience. *Educational Technology Research and Development*, 67(5), 1259–1272. https://doi.org/10.1007/s11423-019-09683-2

Tuszyńska-Bogucka, W., Kwiatkowski, B., Chmielewska, M., Dzieńkowski, M., Kocki, W., Pełka, J., Przesmycka, N., Bogucki, J. and Galkowski, D. (2019). The effects of interior design on wellness—Eye tracking analysis in determining emotional experience of architectural space. A survey on a group of volunteers from the Lublin Region, Eastern Poland. *Annals of Agricultural and Environmental Medicine*. https://doi.org/10.26444/aaem/106233

Vallet, F., Eynard, B., Millet, D., Mahut, S.G., Tyl, B. and Bertoluci, G. (2013). Using eco-design tools: An overview of experts' practices. *Design Studies*, 34(3), 345–377. http://www.scopus.com/inward/record.url?eid=2-s2.0-84875833513&partnerID=40&md5=59b7e2c072134ddd434ecb273450a3e4

Visser, W. (2004). *Dynamic Aspects of Design Cognition: Elements for a Cognitive Model of Design* [Research report]. RR-5144, INRIA. https://hal.inria.fr/inria-00071439/document

Wallach, M.A. and Kogan, N. (1965). *Modes of Thinking in Young Children: A Study of the Creativity Intelligence Distinction.* Holt, Rinehart and Winston.

Wang, W., Duffy, A., Boyle, I. and Whitfield, R. (2013a). Creation dependencies of evolutionary artefact and design process knowledge. *Journal of Engineering Design*, 24(9), 681–710. https://doi.org/10.1080/09544828.2013.825103

Wang, W., Duffy, A., Boyle, I. and Whitfield, R.I. (2013b). A critical realism view of design artefact knowledge. *Journal of Design Research*, 11(3), 243–262. https://doi.org/10.1504/JDR.2013.056591

Wang, W.L., Shih, S.G. and Chien, S.F. (2010). A 'knowledge trading game' for collaborative design learning in an architectural design studio. *International Journal of Technology and Design Education*, 20(4), 433–451. https://doi.org/10.1007/s10798-009-9091-y

Weber, R., Choi, Y. and Stark, L. (2002). The impact of formal properties on eye movement during the perception of architecture. *Journal of Architectural Planning and Research*, 19(1), 57–68.

Weisberg, R.W. (1993). *Creativity: Beyond the Myth of Genius.* W.H. Freeman. http://books.google.com/books?id=_VNwQgAACAAJ

Welch, M. (1998). Students' use of three-dimensional modelling while designing and making a solution to a technological problem. *International Journal of Technology and Design Education*, 8(3), 241–260. https://doi.org/10.1023/A:1008802927817

Welch, M., Barlex, D. and Lim, H.S. (2000). Sketching: Friend or foe to the novice designer? *International Journal of Technology and Design Education*, 10(2), 125–148. https://doi.org/10.1023/A:1008991319644

Wells, J., Lammi, M., Gero, J., Grubbs, M.E., Paretti, M. and Williams, C. (2016). Characterizing design cognition of high school students: Initial analyses comparing those with and without pre-engineering experiences. *Journal of Technology Education*, 27(2), 78–91. https://doi.org/10.21061/jte.v27i2.a.5

Wiggins, G.A. (2006). A preliminary framework for description, analysis and comparison of creative systems. *Knowledge-Based Systems*, 19(7), 449–458. https://doi.org/10.1016/j.knosys.2006.04.009

Wing, J. (2008). Computational thinking and thinking about computing. *Philosophical Transactions of the Royal Society A*. 366(1881), 3717–3725. https://doi.org/10.1098/rsta.2008.0118https://doi.org/10.1098/rsta.2008.0118

Won, P.H. (2001). The comparison between visual thinking using computer and conventional media in the concept generation stages of design. *Automation in Construction*, 10(3), 319–325. https://doi.org/10.1016/s0926-5805(00)00048-0

Woodbury, R. (2010). *Elements of Parametric Design.* Routledge. http://books.google.com.au/books?id=HIM3QAAACAAJ

Woodbury, R.F. and Burrow, A.L. (2006). Whither design space? *Artificial Intelligence for Engineering Design, Analysis and Manufacturing*, 20(2), 63–82.

Wu, Z. and Duffy, A.H.B. (2004). Modeling collective learning in design. *Artificial Intelligence for Engineering Design, Analysis and Manufacturing: AIEDAM*, 18(4), 289–313. https://doi.org/10.1017/S0890060404040193

Yang, E.K. and Lee, J.H. (2020). Cognitive impact of virtual reality sketching on designers' concept generation. *Digital Creativity*. https://doi.org/10.1080/14626268.2020.1726964

Yang, Z., Xiang, W., You, W. and Sun, L. (2020). The influence of distributed collaboration in design processes: An analysis of design activity on information, problem, and solution. *International Journal of Technology and Design Education.* https://doi.org/10.1007/s10798-020-09565-2

Yarbus, A. (1967). *Eye Movements and Vision*. Springer.

Yu, R. and Gero, J.S. (2016). An empirical basis for the use of design patterns by architects in parametric design. *International Journal of Architectural Computing*, 14(3), 289–302. https://doi.org/10.1177/1478077116663351

Yu, R., Gu, N. and Ostwald, M. (2013). Comparing designers' problem-solving behavior in a parametric design environment and a geometric modeling environment. *Buildings*, 3(3), 621–638. https://doi.org/10.3390/buildings3030621

Yu, R., Gu, N., Ostwald, M. and Gero, J.S. (2014). Empirical support for problem-solution coevolution in a parametric design environment [Article]. *Artificial Intelligence for Engineering Design, Analysis and Manufacturing: AIEDAM*, 29(1), 33–44. https://doi.org/10.1017/S0890060414000316

Yu, R., Gu, N., Ostwald, M. and Gero, J. (2015). Empirical support for problem-solution co-evolution in a parametric design environment. *Artificial Intelligence for Engineering Design, Analysis, and Manufacturing: AIEDAM*, 29(01), 33–44. https://doi.org/10.1017/S0890060414000316

Yu, R., Gu, N. and Ostwald, M. (2018). Evaluating creativity in parametric design environments and geometric modelling environments. *Architectural Science Review*, 61(6), 443–453. https://doi.org/10.1080/00038628.2018.1512043

Cognitive Impacts and Computational Design Environments

4.1. Introduction

It is now broadly accepted that the technology employed in a computational design process can potentially shape a designer's thought processes and behaviours (Chien and Yeh, 2012; Mitchell, 2003). The evolution of tools and technology, for example, from chalk to pencil or pen, may not elicit such a change, but the evolution from sketching to scripting probably will. There is, however, another side to this proposition, relating to the larger setting or milieu in which the design process occurs. This setting is made up of an integrated combination of tools, techniques and systems that collectively comprise the environment or ecosystem of the designer. For example, a Traditional Design Environment (TDE) involves particular tools (pencil, paper, eraser), techniques (perspective grids, construction lines, framing guides) and systems (scale measures and reference materials). A Parametric Design Environment (PDE) has its own set of tools (keyboard, mouse, software, computer), techniques (scripting, rendering, generating) and systems (libraries and "apps"). Each of these environments is not just about these elements, it also potentially comprises more than one designer, and a consideration of the communication or interaction that occurs in different types of real and virtual spaces.

The complexity of these design environments, and especially the computational ones, has tended to make them more difficult to study. Nevertheless, speculation about their effectiveness or capacity to support creative processes continues. For example, do computational environments change designers' thought processes and behaviours? Do designers in parametric environments think and operate differently to those in traditional sketching environments? Is it possible to collaborate effectively in a virtual design environment? Can we generate new design instances from a close study and analysis of an existing body of work? To explore these questions, this chapter presents six case studies that examine the impact of different types of design environments on designers and their outputs.

The first case study uses protocol analysis to compare designers' cognitive behaviour in a PDE and in a more traditional, CAD-based Geometric Modelling Environment (GME). In recent years, parametric design has been increasingly used in the architectural profession, however there is a lack of empirical evidence to support a deeper understanding of how PDEs shape designers' ways of thinking or acting. To address this issue, in this case study eight professional architects participated in an experiment wherein each was required to complete two design sessions with design tasks of similar complexity, one in a PDE and the other in a GME. The collected data from the 16 experiments were then used to analyse the cognitive and behavioural differences between the groups in each environment.

The second case study also uses protocol analysis, but in this case the experiment examined the way technology supports synchronous design collaboration. The three design environments compared in this experiment were two variations of synchronous collaborative TDEs – a co-located, pen and paper sketching environment, a remote sketching environment using smart technology – and a 3D Virtual Design Environment (VDE) for remote collaboration. For this study, four pairs of professional architects collaborated on design tasks with a similar level of complexity in each of the three environments. The results of the study were used to evaluate synchronous design collaboration in terms of cognition, communication and interaction.

The third case study presents a parallel exploration of eye movement and the design process in experiments undertaken by two architecture students working in a GME. The students each completed an architectural design task using CAD to collect cognitive data using protocol analysis, while their eye movements were captured using eye-tracking technology. The session was segmented and coded, providing a unique way of considering design cognition and behaviour.

The next two case studies are concerned with design analysis and generation. While in recent years both processes have been enabled using software, it is more common to view generative design as its own type of environment, where multiple combinations of software tools and assessment systems are used to produce an output. The first of these studies revisits grammatical approaches to generative design, and presents a conceptual framework for a Generative Design Grammar (GDG). The framework outlines the general structure of the GDG and its basic components. Integrated with a computational agent model (Russell and Norvig, 2020) and using a VDE as the simulation engine, the GDG can support an intelligent approach towards dynamic and autonomous design.

In this, the fourth case study, a design for a virtual gallery is presented, where a computational agent reasons, dynamically and autonomously generates, simulates and modifies designs. Through this case study the GDG is explained and its potential is explored.

Two garden studies comprise the fifth case. The first demonstrates a multi-component Generative Design Environment (GDE) for analysing and extracting key characteristics from existing designs of a typical style, then generating new works based on those rules. As an example of a GDE, this case analyses and then replicates selected socio-spatial characteristics and aesthetic properties of traditional Chinese private gardens (TCPGs). In the first stage of this study the properties of these complex historic spaces are mathematically derived using connectivity analysis, a variation of a space syntax technique. The data developed through this process is then used to shape the rules of a parametric system to generate new designs with similar spatial properties and structures. Then fractal analysis is used to test if the newly generated garden plans have similar aesthetic properties to the originals. Through this three-stage process (syntactical derivation, parametric generation and fractal dimension analysis) the GDE demonstrates an environment for capturing spatial and aesthetic properties in a parametric system and new tools for design in the context of specific historical sites and approaches. The second study explores Shanghai's famous sixteenth century Yuyuan Garden (also known as the "Yu Garden") using a computational methodology to measure and analyse specific "permeable" and visual properties that have been noted in past research. These spatial properties in the planning of the Yuyuan garden are associated with perceptions of "mystery" and "transparency". The computational methods used to examine these properties are drawn from space syntax (Hillier and Hanson, 1984; Bafna, 2003; Hillier and Kali, 2006; Ostwald, 2011). The comparative analysis produces rich data about the spatial and visual properties of the garden. In order to understand the implications of the data, a computational environment is employed to support data visualisation, design analysis and generation.

The last case study is also concerned with PDEs and GDEs, but more specifically with claims that these environments support the production of creative aesthetic solutions (Park and Nicholas, 2010). In PDEs, architects' design processes differ in a variety of ways from those seen in TDEs and GMEs (Woodbury, 2010; Aish, 2005; Yu et al., 2013) and the capacity to produce creative outcomes is fast becoming the major criterion for judging the success of a particular environment. Despite this, the relationship between parametric design and creativity remains contested. The sixth case study presents a cognitive study that evaluates the level of creativity displayed in designs produced in two environments, PDEs and GMEs. The study correlates the results of design outcome evaluations and design process analysis. To achieve this comparison, a combined method of jury evaluation and protocol analysis is adopted. First, eight professional architects were asked to complete two architectural design tasks in a PDE and a GME, and then the design processes they used were recorded and analysed, in accordance with methods described in the first case study in this chapter. Following the experiment, 19 expert jurors evaluated the design outcomes of the experiment against a pre-defined criterion focusing on creativity. The evaluation criteria addressed innovation, usefulness and unexpectedness. Outcomes of the evaluation and comparison were correlated to findings of a series of protocol studies on related design processes.

Collectively, the six studies reported in this chapter examine design in six distinct environments (TDE, VDE, GME, PDE, GDG and GDE) and two variations of synchronous collaborative TDEs. These studies involved experiments analysing the responses of almost 50 participants, using protocol analysis, eye-tracking and statistical analysis, developing rich and deep data about design cognition and behaviour.

4.2. Case study 1: Designers' behaviour in parametric and geometric design environments

While parametric design has been the subject of growing industrial application in recent years, and research has examined designers' behaviours in parametric environments, there is a lack of evidence to explain *if* or *how* parametric environments affect designers' ways of thinking. In response, this first study compares design cognition in Parametric Design Environments (PDEs) with that in Geometric Modelling Environments (GMEs). For this study, eight architects were recruited to complete two conceptual architectural design tasks using, respectively, a parametric design tool and a geometric modelling tool. Their design processes were video- and audio-recorded as primary data. In the second step, protocol analysis was applied to these recordings to identify the designers' behavioural patterns in both the PDE and the GME. The primary data was transcribed, segmented and categorised according to a coding scheme that was adapted for capturing both CAD and parametric design processes. By comparing the behavioural patterns in the two environments, the characteristics of parametric design are explored and discussed.

Computational design environments – PDE and GME

For comparative purposes, in this study Rhino was chosen as a typical or common GME (Figure 4.1) and Grasshopper as the PDE (Figure 4.2). Grasshopper is an advanced PDE environment for facilitating conceptual design and it has a comparatively wider use in the architectural profession.

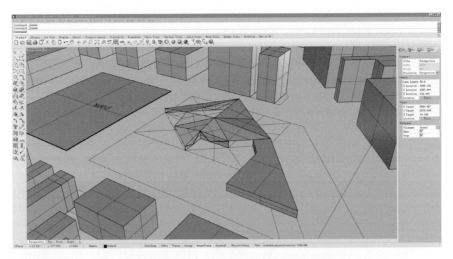

Figure 4.1. Design environment – GME.

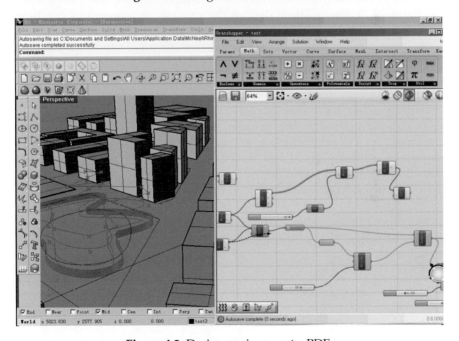

Figure 4.2. Design environment – PDE.

Protocol coding scheme – FBS ontology

In this study, the FBS ontology provides the basis for the coding scheme in the protocol analysis. The FBS ontology (Gero, 1990) was initially formulated as a universal coding scheme for design processes (Kan and

Gero, 2009) and it has been applied in many cognitive studies (Gero and Tang, 1999; Kan and Gero, 2005; Kan and Gero, 2009). Researchers argue that it is potentially capable of capturing most of the meaningful cognitive content of a design process using its system for coding and the transitions between design issues into eight processes (Gero, 1990). The FBS ontology (Figure 4.3.) identifies three classes of variables: *Function (F)*, *Behaviour (B)* and *Structure (S)*. The first, *function (F)*, represents design intentions or purposes; *behaviour (B)* represents the object's derived behaviour *(Bs)* or *expected behaviour from the structure (Be)*; and *structure* (S) represents the components that make up an artefact and their relationships. The coding scheme includes two additional elements that can be expressed in terms of FBS and therefore do not require an extension of the ontology. They are *requirements (R)* and *description (D)*. The first of these encapsulates externally defined or applied requirements and the second, descriptions, refers to the documentation of the design.

The eight design processes of the FBS ontology are formulation, analysis, evaluation, synthesis, documentation and reformulation I, II and III (Figure 4.3). Formulation defines the process that generates functions and expected behaviours, thereby setting up expected goals from the requirement, while synthesis generates a structure as a candidate solution. Analysis produces a behaviour from the existing structure and evaluation compares *Bs* and *Be* to determine the success or failure of the candidate solution. Reformulation is the process that arises from the structure revisiting either the behaviour or the function, which is a multi-stage reconstruction or reframing process. Across the eight design processes, the three types of reformulation are the dominant actions that potentially capture creative aspects of designing by introducing new variables or new directions (Kan and Gero, 2008). By calculating the transitions between design issues using empirical data, a range of tests can be conducted.

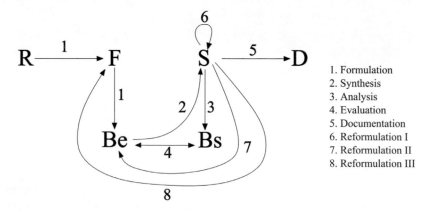

1. Formulation
2. Synthesis
3. Analysis
4. Evaluation
5. Documentation
6. Reformulation I
7. Reformulation II
8. Reformulation III

Figure 4.3. The FBS ontology (after Gero, 1990).

Development of a coding scheme to study designing in PDEs

The explicit application of rule-based algorithms and their associated processes, is the primary means by which traditional GMEs and PDEs can be differentiated, with the latter, utilising algorithmic tools as a core part of the design process. With this distinction in mind, an FBS ontology-based custom coding scheme is required to analyse design in a PDE. This scheme must be able to encode two types of behaviours of parametric designers: their use of design knowledge and their use of rule algorithms. During the design process in GMEs, only design knowledge use occurs, because designers only consider their product, which for an architect, is typically a building. In contrast, in PDEs both aforementioned algorithmic behaviours occur, because designers in PDEs incorporate rule algorithms while producing the final product. To record and deconstruct designers' activities from both types of behavioural perspectives, the primary variables (Be, Bs, F, R, S) are decomposed in terms of two associated design space categories: "rule algorithm space" (denoted as a superscript R), and "design knowledge space" (denoted as a superscript K), as seen in Figure 4.4. The rule algorithm space's structure variable (S^R) can have additional subclasses, relating to specific rule algorithm activities in PDEs, thereby enabling detailed representations of how design processes are supported by parametric tools. For rule algorithm spaces in PDEs, the original FBS ontology requires a modification to employ multiple instances of the FBS variables. Meanwhile, design knowledge spaces can comply with the original unmodified FBS ontology since they are similar to TDEs. Although the FBS model is utilised for the present coding scheme, one of its original codes *description* (D) is excluded, since it is rarely applicable to PDEs. PDE tools automatically describe and document 3D models during the scripting or programming activities of a designer, while the designer is focusing on S. Hence this case study does not include a "D" code.

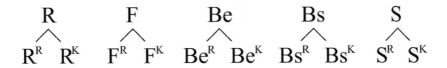

Figure 4.4. Adopting the FBS ontology for studying PDEs.

4.2.1. Research design

Eight designers participated in the experiment, all were professional architects with an average of eight years of experience in a design practice, and with no less than two years' experience using parametric design. These controls were chosen to ensure that the participants were experienced both in architectural design and in using parametric software. Previous protocol studies have tended to select participants with experience levels ranging from five to ten years (Gero and Kannengiesser, 2012; Kan and Gero, 2009); however, most parametric designers tend to come from a younger generation. Therefore, architects with eight years' experience are considered as sufficiently experienced amongst the younger generation and are suitable for the current study.

The experiment consisted of two design sessions: one used Rhino as the GME and the other used Rhino and Grasshopper as the PDE. Designers were given 40 minutes for each design task. Task 1 was a community centre conceptual design and Task 2 was a shopping centre conceptual design, both containing specific functional requirements. A pre-modelled site was provided to the designers. Because the study was focused on exploring designers' behaviour at the conceptual design stage, they were only required to consider concept generation, simple site planning and general function zoning, and no detailed plan layouts were required.

The design sessions and tasks were randomly allocated across the participants. During the experiment, designers were not allowed to sketch manually to ensure that almost all of their actions were recorded in the computer and thus the design process was purely within the PDE and GME. This minimises the impacts of other variables and focuses the outcome on the different design environments. During the experiment, both designers' activities and their verbalisations were video-recorded.

The data segmentation and coding processes used in this study are based on the "one segment one code principle" (Pourmohamadi and Gero, 2011). This means there are no overlapping codes or multiple codes for a segment. If there are multiple codes for a segment, the segment is sub-divided.

4.2.2. Result 1: Design issues and processes

For a comparative analysis of the overall distribution of design issues in the GME and the PDE, boxplot analysis is used to report the relative portion that each design variable takes up in a set of protocols. The boxplot graphically describes the data using five variables (smallest observation, lower quartile, median, upper quartile, and largest observation). Figure 4.5 is the boxplot of the overall design issues distribution in the GME and in the PDE. This figure reveals that the two design environments produce similar distributions of issues. In particular, more cognitive effort is expended on *structure (S)* than any other in both environments; this is followed by *behaviour from structure (Bs)*, *expected behaviour (Be)*, *function (F)*, and the least effort is expended on *requirements (R)*.

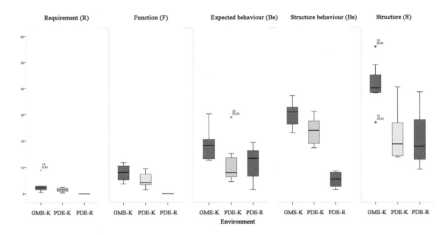

Figure 4.5. Boxplot analysis of overall design issues distribution.

In order to compare the overall distribution of design issues between the GME and the PDE, paired sample T-tests were used. Because the experiment uses the same group of designers in two different environments, the paired sample T-test is a suitable analytical method to construct the comparison (Table 4.1). The results suggest that there is no significant difference in terms of overall distribution of design issues between the GME and the PDE (P>0.05). That is, designers' overall efforts invested on design issues are similar in the two environments. This suggests that whatever digital design tools they used, designers' cognitive activities at the design thinking level (which deals with their design habit, strategy, or preferences) do not vary significantly.

Table 4.1. Paired sample T-test of overall design issues between the GME and the PDE.

GME vs. PDE	Mean	Std. Deviation (SD)	t	Sig. (2-tailed) (P)
R	1.30875	2.38226	1.554	.164
F	−.44000	3.70157	−.336	.747
Be	−1.25375	6.86746	−.516	.622
Bs	1.52125	3.95736	1.087	.313
S	−1.07875	8.38667	−.364	.727

The boxplot of results for overall distribution of design processes in the GME and the PDE shows that more cognitive effort is expended on the analysis and reformulation I processes (Figure 4.6). This is followed by evaluation, synthesis and reformulation II, with the least effort expended on formulation and reformulation III. This may be because designers analyse the existing geometric model/script relatively more frequently, and after the evaluation process, they modify the model/script in response.

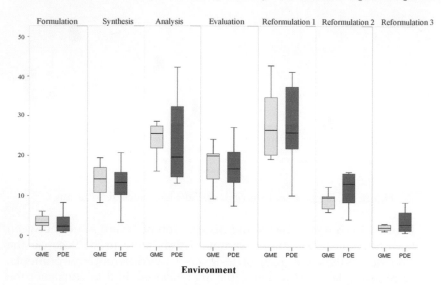

Figure 4.6. Boxplot analysis of overall distribution of syntactic design processes.

With one exception, paired sample T-test results for the data suggest that there is no significant difference in terms of overall design processes between the GME and the PDE (P>0.05) (Table 4.2). The exception is the occurrence of reformulation II in the PDE, which is higher than in the GME. This means that designers in the PDE reformulate *behaviour (Be)* more frequently, using reasoning derived from the existing geometric model or rule, to reset the algorithm goals or the way to achieve them.

Table 4.2. Paired sample T-test of syntactic design processes between the GME and the PDE.

GME vs. PDE	Mean	Std. Deviation (SD)	t	Sig. (2-tailed) (P)
Formulation	.76250	2.27090	.950	.374
Synthesis	1.12500	4.99335	.637	.544
Analysis	.72500	10.54335	.194	.851
Evaluation	.76250	6.10502	.353	.734
Reformulation I	.45000	14.35708	.089	.932
Reformulation II	−2.87500	4.54116	−1.791	.116
Reformulation III	−.95000	3.88109	−.692	.511

Significant qualitative differences can be observed between the data developed from the PDEs and the GMEs. Figure 4.7's boxplot illustrates the distribution of design issues in design knowledge spaces compared to rule algorithm spaces. Based on this distribution, it is apparent that in GMEs more cognitive effort is expended on design knowledge issues compared to in PDEs. The results also indicate that the actual make-up of design issues is dissimilar between the PDEs and GMEs. Despite the overall distribution of design issues being similar, in the PDE some issues related to design knowledge are replaced by those related to rule algorithms. Also notable in these results is the significant impact of rule algorithms on expected behaviour "Be^R". This is potentially caused by designers' frequent reflection on how to achieve rule algorithm related goals in PDEs. Furthermore, the impact of rule algorithm on structure "S^R" is also visible, possibly arising from the PDEs practice wherein designers consider the structure of design in parallel with the rule algorithm structure.

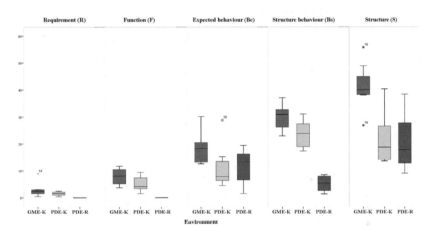

Figure 4.7. Boxplot analysis in GMEs and PDEs of the distribution of design issues in design knowledge spaces and rule algorithm spaces.

4.2.3. Result 2: Designers' cognitive effort

Cumulative analysis was utilised for the aggregation process of design issues, to illustrate the expended efforts of designers during each design session. Such analysis can identify the total cognitive effort expended on each design variable during the design process. Cumulative occurrence per design issue can be calculated by: the number of design issues within a single category (that occurred thus far in a session), divided by total number of design issues of that specific type (Gero and Kannengiesser, 2014). Equation (1) defines this.

$$\text{Cumulative issue} = \frac{\text{The number of issues that have been coded so far}}{\text{The total number of design issues of this category}} \quad 100\%$$

The overall relative effort expended across the two levels of design activities is illustrated in Figure 4.8. The vertical and horizontal axes represent the average values of the relative effort of the eight protocols and the normalised segment numbers, respectively. The total distribution of cognitive effort depicted in Figure 4.8 suggests that initially it is invested in design knowledge rather than rule algorithm. As the design session proceeds however, the cognitive effort expended on design knowledge drops from 100% to approximately 60% of the total, in a trend broadly reminiscent of a "decay curve". In parallel, as the design session proceeds, the cognitive effort expended on rule algorithms increases from 0% to approximately 40% of the total, broadly following an "excitation curve". Therefore, it might be inferred that in parametric design, designers invest the most cognitive effort in design knowledge, with parametric scripting mainly being used to support the generation of a design model. Furthermore, the designers often commenced their sessions by considering design knowledge-related issues, such as building functions. As the process proceeded however, they gradually increased their cognitive investment in parametric scripting. In the following sections these results are analysed in more detail, to consider cognitive behaviours related to *Be, Bs* and *S*, and draw some observations from the results.

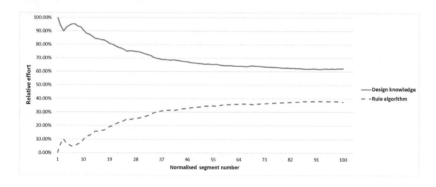

Figure 4.8. Overall relative cognitive effort.

Be-related effort for the two categories

Expected behaviour, "Be" can be described as behaviour where designers either apply learnt or experienced knowledge to "speculate what effect could fulfil a purpose" prior to suggesting a specific structure (Jiang, 2012, pp. 36-37). As such, in terms of rule algorithm related activities, "Be^R" reflects the ways designers devise rule algorithm related goals or think about possible options for achieving such goals. Figure 4.9 illustrates the relative cognitive effort expended during parametric design processes on expected behaviour (Be) within each of the two categories. These efforts within the design knowledge category (Be^K) start with high values, which then decrease as the design session progresses. Conversely, efforts expended on expected behaviour (Be) at the rule algorithm level (Be^R), increase towards the end of a design session. At the conclusion of the design session, the cognitive effort related to rule algorithms slightly exceeds that for related design knowledge. This could be read to suggest that designers focus more on exploring rule algorithm approaches to reach their objectives, towards the later part of their design session.

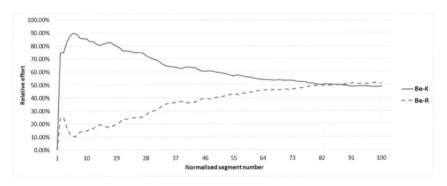

Figure 4.9. Relative cognitive effort expended on Be within two categories.

Bs-related effort for the two categories

Behaviour derived from structure "Bs" signifies, in the context of design knowledge related activities "Bs^K", how designers evaluate existing structure and/or geometry whereas "Bs^R" signifies how designers evaluate the rule algorithms' structure. Figure 4.10 illustrates the relative cognitive effort expended during parametric design processes on structure behaviour (Bs) within each of the two categories. During the duration of a design session, the efforts expended on structure behaviour within the design knowledge category (Bs^K) exceed the efforts expended on it within the rule algorithm category (Bs^R). There is a reduction (from 100% to near 70%) of activities associated with design knowledge during the first one-third of a design session, then it remains relatively consistent (around the 70%–80% mark). Correspondingly, there is an increase (from 0% to near 30%) of activities associated with rule algorithms during the first third of a design session, and then that remains steady (around the 20%–30% mark). Figure 4.10 also demonstrates that rule algorithm related activities "Bs^R" do not commence at the start of a design session. A potential explanation for this could be that designers may not wish to adjust their rule algorithms, until the moment in their design session when their initial concept has reached a mature stage.

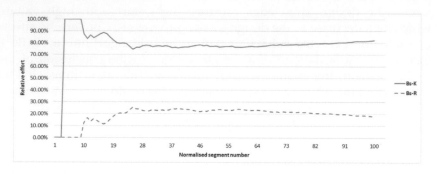

Figure 4.10. Relative cognitive effort expended on Bs within two categories.

S-related effort for the two categories

Structure "S" can be defined as the components of the design product and their relationships, which describe "what it is" (Gero and Kannengiesser, 2004, p 374). In relation to rule algorithm related activities "S^R" describes components of rules and their relationships for parameterisation. In relation to design knowledge related activities, "S^K" describes geometric elements and/or their relationships in the design. Figure 4.11 presents the relative cognitive effort expended during parametric design processes on structure (S) within each of the two categories. The efforts expended on structure (S) within the design knowledge category (S^K) start at 100% at the commencement of sessions and then decrease to around half as the design session progresses to its conclusion. Conversely, efforts expended on structure (S) at the rule algorithm level (S^R) increase from 0% initially, to near 50% towards the end of the design session.

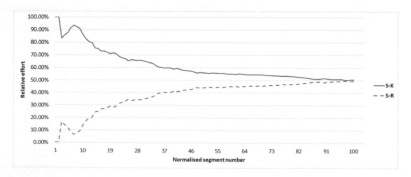

Figure 4.11. Relative cognitive effort expended on S within two categories.

The results of the analysis of cognitive effort are indicative of the usefulness of this dual category – design knowledge and rule algorithms – for deepening our understanding of the dominant behaviours that occur when designing in PDEs. Design knowledge associated cognitive effort changes from 100% to 60%, from the start of the design session towards the end, and conversely, rule algorithm associated cognitive effort changes from 0% to 40%, from the start of the design session towards the end. These results show that parametric design processes are cognitively dominated by design knowledge associated activities. If such results are more broadly generalisable, then it suggests that experienced practicing architects utilise the algorithmic tools of PDEs more frequently over the course of their design session. In other words, during the design process, architects substitute cognitive investment in design knowledge with investment in rule algorithms. Such increases, nevertheless, are at the expense of cognitive effort associated with design knowledge. This interchange between design knowledge and rule algorithms may point to an approach for encoding design patterns in PDEs – which in turn could lead to modularised and reusable rules or rule-sets unique to a designer's own preference or style of designing – and potentially encapsulate a language or style of designs. Such an approach could generate individual designs (uniquely generated through parametric design) using encoded design patterns that may respond to design requirements, or individual designer's (or even design teams if the encoded patterns were shared) design style, as well as variations within that style. Design patterns of this type will be discussed in greater detail in the following section.

4.2.4. Result 3: Design patterns

Markov model analysis of the data was conducted to calculate the transition probabilities between design issues. A Markov model describes the probabilities of moving from one state to another (Ching and Ng, 2006; Meyn and Tweedie, 2009). In the present context it is used as a quantitative tool to study design activities based on the transition probabilities in the FBS ontology between design issues and design processes. To demonstrate the main activities of designers in the present study, transitions with a probability larger than 0.4 were selected (Figure 4.12). During the trial analysis of the data a threshold of 0.3 was initially set, but the Markov model filtered with this threshold contains too many transitions to focus and prioritise on, and therefore 0.4 was used as the threshold. The results of the Markov analysis of the data show that the transition probabilities are very similar between the GME and the PDE (Figure 4.12). The main difference between the GME and the PDE is that the transition probability from F to S is much higher in the PDE.

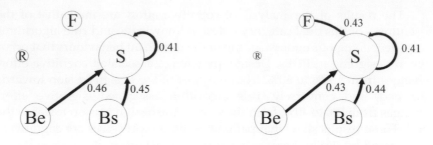

Figure 4.12. Left: main transitions of the 1st order Markov model in the GME. Right: main transitions of the 1st order Markov model in the PDE.

The result for transition probability from *F* to *S* is interesting because within the FBS ontology it is not regarded as a routine path. Nevertheless, previous research into software designers' behaviour indicates that *F* to *S* is a typical transition (Kan et al., 2013) which has now been identified in the PDE. During the F to S process, designers select an existing structure/ solution for the particular function/design problem, based on their experience or knowledge, which is a process of picking up an existing design pattern for the problem. Since software designers use design patterns when programming and scripting (Fowler, 2003; Gamma et al., 2002), it could be inferred that when architects apply programming and scripting in their design process in a PDE, they exhibit a similar use of design patterns. A design pattern is an important concept in both architectural design and software design. In software design, it assists programmers to work more efficiently and it makes the scripting process traceable. In the PDE, generalising useful design patterns assists architects in their scripting process.

The idea of a design pattern was initially popularised by Christopher Alexander, who stated that "each pattern describes a problem which occurs over and over again in our environment, and then describes the core of the solution to that problem, in such a way that you can use this solution a million times over, without ever doing it the same way twice" (Alexander et al., 1977, p. x). That is, a pattern is the documentation of a solution suitable for certain kinds of design problems that may occur frequently (Dawes and Ostwald, 2017; Dawes and Ostwald, 2020). Woodbury et al. observe that patterns "are a way to identify successful general strategies that exemplify a key concept in a memorable fashion that can easily be taught" (Woodbury et al., 2007, p. 299). Patterns come from designers' experiences (Fowler, 2003), which can be seen as an "induction" process, whereby designers generalise samples from their own experience or from observation of other designers, and abstract the problem-solution pair and formalise "patterns" that can be re-used (Razzouk and Shute, 2012). These patterns can be improved or combined into a network of connections depending on the design purpose (Alexander, 1979). The pattern itself is an abstract formulation that becomes specific when designers revise and apply it in a particular context (Alexander et al., 1977).

In the results of the Markov analysis, there are more *F-S* design patterns in the PDE than in the GME. Parametric design is a combination of architectural design and rule algorithmic design (Yu et al., 2013). From observation of the results of this study, when participants design in the PDE, three types of design patterns repeatedly occur (Figure 4.13).

- Design pattern 1 occurs at the design knowledge level, which captures architectural design solutions that serve design functions. This type of pattern is similar to those illustrated in Alexander's theory (Alexander et al., 1977).
- Design pattern 2 occurs across two levels of activities, which are scripted structural solutions serving design functions. This type of design pattern is unique in the parametric design process.
- Design pattern 3 occurs at the rule algorithm level and it captures scripted structural solutions serving script functions. This last type of design pattern complies with Woodbury's parametric design patterns (Woodbury, 2010; Woodbury et al., 2007).

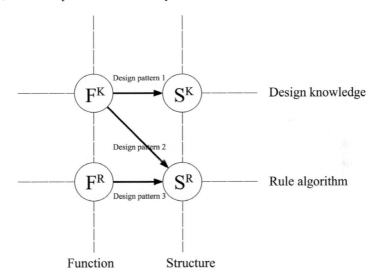

Figure 4.13. Three types of design patterns in the PDE.

4.2.5. Result 4: Co-evolution process in parametric design

To find their preferred approach or solution from among a myriad of potential alternative solutions, designers continually re-define or reframe their own design intentions in an iterative series of activities during the process of designing. This type of iteration is not merely a cyclical series of events, since it necessitates designers re-evaluating their assumptions, problems and solutions during the process, which consequently leads to evolution and changes in the design parameters. Such advanced design processes that are used by experts, involve designers reviewing their representations of design problems and solutions in a kind of interactive flux (Cross, 2011). A research approach seeking to understand designers' problem-solution co-evolution processes is therefore both necessary and useful, for calculating and visualising the transitions that occur between the problem space and the solution space.

The discontinuity ratio within PDEs and GMEs

This section introduces the quantitative methodology used for exploring the co-evolution of problem and solution processes in design. The frequency of occurrence of transitions between these spaces provides a robust indicator of the co-evolution process. The frequency of occurrence is calculated for these processes' discontinuity ratio, which is defined as the ratio between number of transitions and the total number of segments, as shown in equation (2). The ratio can reveal the frequency of occurrence of transitions between problem space and solution space, within a designated period. A higher frequency of transitions will result in a higher discontinuity ratio.

$$\text{Discontinuity ratio} = \frac{\Sigma \text{ Transition number}}{\Sigma \text{ Overall segments number}} \times 100\% \qquad (2)$$

A comparison of the discontinuity ratio in GMEs and PDEs can be seen in Figure 4.14, where each of the ten sub-sessions, depicted along the horizontal axis, are equally divided segments of the entire design session. The GME discontinuity ratio data in the figure shows a relatively flat or even rate of transition between problem and solution spaces across the

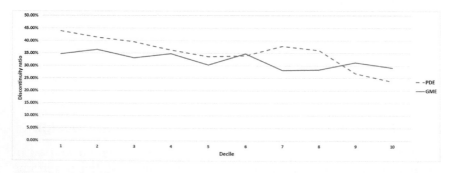

Figure 4.14. Problem and solution space discontinuity ratio in GMEs and PDEs.

entire design session. By contrast, the PDE discontinuity ratio is more variable, starting off high, decreasing for a while, then briefly rising and finally falling again towards the end of the design session. The sub-sessions near the middle and end of the design session (segments 6 to 8) reveal a larger discontinuity ratio in the PDE compared to the GME. This result may potentially correspond to the phase of the session where designers are most actively engaging in the co-evolution process.

Analysis of co-evolution processes within PDEs vs. GMEs

It is suggested in the work of Simon (1973) that designers constantly reformulate their design problems by repeatedly returning to the design problem space. Designers then develop a design to their satisfaction, by way of an ongoing co-evolution between the design solution and problem spaces (Maher and Tang, 2003). Hence the co-evolution of these two design spaces offers an important perspective from which this study examines designers' behaviour in GMEs and PDEs. Using discontinuity ratio, this study calculates and measures transitions in the co-evolution process in GMEs and PDEs. Transitions are important because they are associated with design creativity (Dorst and Cross, 2001). Parametric design, however, transcends traditional GME-based thinking, by visualising the conceptual ideas of the designer as algorithms to support the definition and modification of design parameters and parametric relationships, which expands design thinking. In PDEs, it is observed that more frequent interactions occur between the computer and the designer, since the designer must switch between the two interfaces – a scripting interface and a geometric interface – during a design session, in order to facilitate effective flow of information. In PDEs, designers operate in a recursive circular process, becoming both informed and inspired by their on-screen visuals, which they then use as the basis for new rule creations or modifications. This process continuously co-evolves – building upon the co-evolution of the design solution space and the design problem space – thereby the design progresses until the outcome has been finalised. Such a process has been described as a "seeing-moving-seeing" response to instant feedback received from the execution of design actions (Schön, 1992; Goldschmidt and Porter, 2004). In essence, the designer in a PDE sees what is on the screen through the geometric interface and is informed by the visuals. Then, based on these representations, the designers make adjustments to refine their models and scripts, before reviewing the outputs and their potential influence. This recursive, circular process, then begins anew, with the most recent outputs as the new inputs. Whereas equivalent recursive processes may occur in TDEs, the computational power at the heart of the PDE makes this process occur more rapidly.

The findings in this case suggest that designers in GMEs and PDEs both invest similar overall efforts into their design problem and solution spaces, and in both environments they tend to commence their design sessions (i.e. during the initial analysis of their design brief) with problem-driven tactics. Therefore, in each design environment they exhibit a greater focus on the problem space at the start of their design sessions. The analysis then suggests that a key difference among designers' behaviour in GMEs, compared to that in PDEs, occurs in the next part of the early design stage. In PDEs, designers typically move to solution spaces at this time, as they develop design parameters and rule algorithms. In contrast, in GMEs designers are still often focused on their problem space. Past research suggests that design creativity is enabled by solution-driven design (Kruger and Cross, 2006), and this may imply that PDEs are more likely to inspire design creativity at the early design stages, compared to GMEs. The measurements of cognitive transitions in this case between problem and solution spaces (i.e. calculated discontinuity ratio), also indicate a broad similarity between GMEs and PDEs as discussed previously. But looking more closely at the three stages of a design session, a key difference in discontinuity ratio is found between GMEs and PDEs in the early design stage. During that early stage there are significantly less transitions in the GME than in the PDE, which is perhaps indicative of PDEs supporting a stronger co-evolution process. Since prior studies (i.e. Zeiler et al., 2007) have suggested that early design stages are where a majority of key design decisions are made, formulating solution development theories and approaches early is beneficial to solution optimisation (Cross, 2004), and it follows that PDEs may present a better opportunity for designers to create more thoroughly considered design solutions. This is because there is a greater frequency of interactions at the early design stage in the PDE compared to the GME.

Rule algorithmic effects on co-evolution processes in PDEs

Working from the perspective of the two levels of activities in PDEs, and based on Jiang et al.'s division of problem-solution related design issues (Table 4.3), both problem-related and solution-related design issues in PDEs can be further categorised into: design knowledge level's problem-related issues (P_K) (including R^K, F^K, Be^K), rule algorithm level's problem-related issues (R_R) (including Be^R), design knowledge level's solution-related issues (S_K) (including Bs^K, S^K), and rule algorithm level's solution-related issues (S_R) (including Bs^R, S^R), as described in Table 4.4. Since it is an inherent part of the parametric design process for designers to explore problems and solutions via rule algorithms (in addition to considering them from a traditional design knowledge perspective), it makes sense to divide design issues into the aforementioned groupings.

Table 4.3. Mapping FBS design issues onto the problem and solution spaces (Jiang et al., 2014).

Problem/solution space	Design issue
Reasoning about problem	Requirement (R)
	Function (F)
	Expected behaviour (Be)
Reasoning about solution	Behaviour derived from structure (Bs)
	Structure (S)

Table 4.4. FBS design issue mapping to problem/solution spaces in PDEs.

Problem space/solution space	Design issue mapping
Design knowledge level reasoning about problems (P_K)	Requirement (R^K)
	Function (F^K)
	Expected behaviour (Be^K)
Rule algorithm level reasoning about problems (P_R)	Expected behaviour (Be^R)
Design knowledge level reasoning about solutions (S_K)	Behaviour derived from structure (Bs^K)
	Structure (S^K)
Rule algorithm level reasoning about solutions (S_R)	Behaviour derived from structure (Bs^R)
	Structure (S^R)

Figure 4.15 shows the distribution for the discontinuity ratio of each PDE transition, illustrating the design sessions' eight transition types between problem and solution spaces. The horizontal axis of the figure charts each decile (of ten sub-sections within a session). The vertical axis shows the eight participants' average transition pattern discontinuity ratio per decile. The following three time-based descriptions are used as indicators of progress towards completion of a design: the period known as "early design stage" being 1.0 to 3.3 on the horizontal axis, "mid design stage" as 3.4 to 6.7, and "end design stage" as 6.7 to 10. The eight lines in Figure 4.15 illustrate the following types of transitions between design problem and design solution spaces: P_R to S_R, P_K to S_K, P_R to S_K, P_K to S_R, S_R to P_R, S_K to P_K, S_K to P_R and S_R to P_K. The prevalent transition in the early design stage is P_K to S_K, and in the mid design stage it is P_R to S_R (with the transition from S_K to P_K still active to some extent), and in the end design stage more transitions are between P_R and S_R. This indicates that the co-evolution process focuses on knowledge level at the start of the design session, is active at both knowledge and rule algorithm levels at the mid design stage, and focuses on the level at the end of the design session. A possible explanation for this pattern may be that at the start of a session designers initially consider the design brief by applying their design knowledge (akin to most common architectural practices), and then only

later explore their concepts and goals using rule algorithm processes. Designers continually redefine and reinterpret their design problem as they explore solutions at the design knowledge level. Additional experimental observations noted that designers had a tendency to start off using their design brief document, then analysed the site to develop their initial concepts, before considering the structural forms of their design and setting parametric rules to explore ways to implement their design goals. A further observation is that during sessions designers employed a gradually evolving conceptual development approach, that repeatedly returned to the design knowledge level to evaluate their latest evolving design form. Finally as the completion of the design session drew near, designers focused on rule algorithm design, using parametric rules to complete their geometric design modelling.

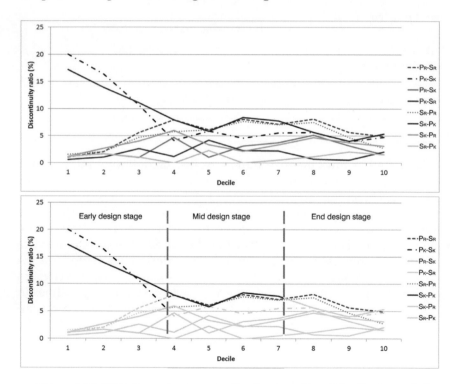

Figure 4.15. PDE discontinuity ratio distributions. Top: Distributions of all transitions. Bottom: Each design stage's main transitions.

Co-evolution process model for PDEs

To synthesize the findings developed through this case, Figure 4.16 illustrates a model of the co-evolution process in the PDE; and illustrates the transitions between design problem and design solution spaces, for both the knowledge and rule algorithm levels. The evolution of the problem space (P) from time "t" to time "t+1" is indicated in the figure by lateral moves horizontally, while vertical moves in the figure suggest processes where "the solution refocuses the problem" and "the problem leads to the solution" (Maher and Poon, 1996). This conforms with Maher's and Kundu's (1993) argument that a solution space S(t) results in new requirements in P(t+1) which were not present in the original problem space P(t), i.e. that solution spaces S(t) are not only spaces where design solutions are explored, but rather that design requirements also change along with design solutions. The model of the co-evolution process in the PDE developed through the results in the present chapter is detailed in Figures 4.17, 4.18, and 4.19.

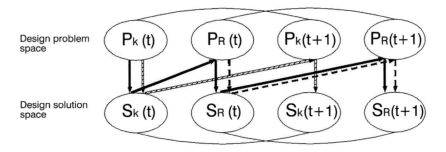

Figure 4.16. The PDE co-evolution process model as identified by findings of this study.

Figure 4.17 illustrates co-evolution on the rule algorithm level between design problem and design solution spaces, as indicated by dashed arrows (a frequently occurring process in PDEs). Solutions are explored by designers to meet rule algorithm requirements or goals, P_R (t), from solution space S_R (t), to add/refine new requirements, thereby reformulating rule algorithm problem P_R (t+1).

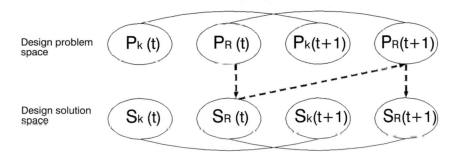

Figure 4.17. Rule algorithm level's co-evolution processes.

Figure 4.18 illustrates co-evolution on the design knowledge level between design problem and design solution spaces, as indicated by solid dashed arrows. This is a common process in PDEs, and an aspect of designers' behaviour which has similarities with processes in TDEs (Maher and Poon, 1996; Dorst and Cross, 2001).

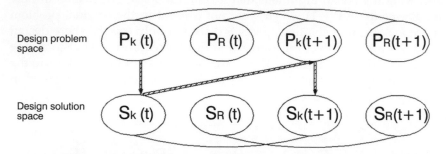

Figure 4.18. Design knowledge level's co-evolution processes.

Figure 4.19 presents the model of co-evolution on both the design knowledge level and the rule algorithm level, as indicated by solid arrows. Designers during this co-evolution process start from the problem space on the design knowledge level P_K (t), then during exploration in the solution space S_K (t) new requirements emerge on the rule algorithm level (which refined the design problem space on the rule algorithm level P_R (t)). That is, exploration of the problem and solution alters the direction on the rule algorithm level, which is a PDE process in which designers' exploration of design solutions leads to redefining and reformulating the design problem on both the rule algorithm and design knowledge levels.

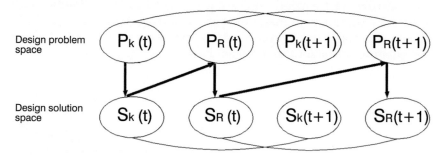

Figure 4.19. The co-evolution process across the design knowledge level and the rule algorithm level.

During the act of designing, designers iteratively redefine and reformulate the problems they are trying to solve, which leads them to repeatedly return to the design problem space (Simon, 1973). Through this behaviour, their design progresses via cognitive activities that switch between design problem and design solution spaces, until they finally reach a "satisfactory" outcome (Maher and Tang, 2003). Since the concept of "design" in itself can be defined as a special type of problem solving process (Simon, 1969) in which the problems to be solved may be ill-defined (Simon, 1973; Corne et al., 1993) or unclearly defined (Maher et al., 1996; Chi, 1997), it follows Cross' (2011) and Schön's (1983) suggestions that creative design processes are a kind of exploration process during which solution spaces and problem spaces intermingle and unstably evolve towards some "emergent bridge" that is deemed satisfactory. Understanding the co-evolution process is key to creative design processes. The primary role of rule algorithms (as they operate in parallel with traditional design means) in parametric design, is to give PDEs a fundamentally different design approach compared to the more traditional design in GMEs (Yu, Gero et al., 2013). Case Study 1 has briefly examined the co-evolution process in PDEs, supported by empirical data obtained from experiments involving professional designers. Results of the experiments showed that utilising a problem-solution dichotomy was sufficiently comprehensive and effective in capturing the behaviours of parametric designers, to increase our understanding of the PDE design process. By dividing design activities into two levels (design knowledge level and rule algorithm level) and noting how frequently transitions occurred between the problem and solution spaces, the following key characteristics of the PDE co-evolution process have been identified. The first characteristic is that transitions across design knowledge and rule algorithm levels only occur relatively infrequently; the designers' co-evolution process generally takes place within the design knowledge or rule algorithm levels. The second characteristic is that the co-evolution process expends more cognitive effort at the rule algorithm level near the end of the design session, and focuses more on the design knowledge level in the early design stage. Therefore the activities that are most representative of the essence of parametric design (i.e. rule-algorithm level activities) appear to be less influential in the early stages of the design session, and more crucial to the design process in the later stages. Third, this study identifies three sub-processes within co-evolution at the design knowledge level, the rule algorithm level, and across these two levels.

4.3. Case study 2: Cognitive studies of design collaboration in a virtual environment

In practice, designers collaborate with other designers from their own disciplines, as well as with experts from other fields. Lahti et al. (2004) define design collaboration as a process where designers dynamically communicate and work together, aiming to collectively establish design goals, search through design problem spaces, determine design constraints and construct design solutions. While individual designers contribute to design development, collaboration implies teamwork, negotiation and shared representations. Most of this teamwork occurs physically in the same space, in meeting rooms or studios and over desks or on shared screens. Increasingly, however, these interactions occur online in shared virtual design environments. These range from real-time multi-accessible BIM models to immersive, interactive VR environments.

This second study in this chapter investigates remote design collaboration in a 3D virtual world. Its goal is to understand changes in design behaviour that occur when designers are physically remote, but virtually co-located as avatars within their design, represented in a 3D model. In order to provide a baseline, the results of the virtual environment are compared with those of two other environments: (1) a co-located sketching environment, and (2) a remote sketching environment. The first comparison aims to better understand designers' activities and behavioural changes when collaborating remotely through 3D virtual worlds in relation to the traditional collaborative scenarios where designers are co-located. The second comparison examines the impact of different modes of design representation in remote design collaboration to see if the use of 2D sketches or 3D models is significant. For this study, four pairs of professional architects collaborated on a different design task with a similar level of complexity in each of the three simulated design environments. The comparison of the same pairs of designers in different environments provides an indication of the different impact of the environments on design cognition.

4.3.1. Collaborative design studies and technologies

A common research focus of past collaborative design studies is communication between participants. The goals of such research includes the analysis of components of collective thinking and team behaviour, and the investigation of social behaviours such as community sense, open participation and level of participants' awareness in the digital media (Lee et al., 2016; Lee et al., 2019; Phare et al., 2016; Phare et al., 2018). Collaborative design technologies support a team of designers to work together on a design project while either remotely or co-located. Since the communication focuses on shared design representation (sketches, drawings and models), the important features of collaborative design technologies are those that support the creation, modification and exchange of shared drawings or models. It is therefore assumed that team design cognition, communication and interaction have changed with the introduction of these technologies.

In the past two decades, a variety of disciplines have participated in developing, applying and validating tools that are designed to address human collaboration at work; commonly known as Computer Supported Collaborative Work (CSCW) systems. Research in this field commenced in the 1960s with the Stanford Research Institute's work on innovative interaction techniques. This was followed by research at Xerox PARC in the 1970s and 1980s and the development of User-centred Design (Norman and Draper, 1986), which led to research into Human-computer Interaction (HCI). In the 1990s, a branch of HCI evolved into CSCW, becoming a multi-disciplinary field that contributes to the development of collaborative technologies.

The digital media used for representation in collaborative environments are core to their operations. These can include images, sketches, structured graph models, 2D and 3D geometric CAD models and 3D object-oriented models. Two common digital sketching tools for supporting design collaboration are *Group Board* and *Sketchup*, and some major CAD platforms, such as *ArchiCAD*, also have integral teamwork components. The technologies used to support interaction include standard graphical user interfaces (GUI) utilising keyboard and mouse, touch screens and tabletops, AR and gaming 3D motion controllers. Designers can also communicate with each other using text chat, voice over IP or video.

With technical advances in the field and the wider adoption of high-bandwidth internet, a new generation of collaborative technologies has also emerged. Two of these technologies that support synchronous design collaboration are 3D virtual worlds and 3D models. 3D virtual worlds enable designers who are remotely located to be represented as, and to collaborate through, avatars in a virtual environment with the shared presence of both 3D design representations and the design team. Such 3D virtual worlds as "places" on the internet support activities ranging from social interaction, to e-commence, to education and design. One of the main characteristics that distinguish 3D virtual worlds from conventional virtual reality is that 3D virtual worlds allow multiple users to be immersed in the same virtual environment, supporting a shared sense of place and presence (Singhal and Zyda, 1999). This is a key reason for adopting 3D virtual worlds for remote design collaboration. Multi-user 3D virtual worlds grew rapidly in the 2000s, with examples such as Second Life (www.secondlife.com) being experienced by millions of users, providing a strong case for the impact of virtual communities and the activities that take place in them.

3D virtual worlds have been closely associated with design ever since the early conceptual formation of the field (Ostwald 1993, 1997). On one hand, the rich knowledge and design examples in the built environment provide a good starting point for the development of 3D virtual worlds.

On the other hand, creating 3D virtual worlds, especially for the gaming industry, has become a specialist design skill in itself. For the AEC industry, the availability of 3D virtual worlds and high-speed internet have transformed remote design collaboration, allowing designers to collaborate on projects without being physically present. High-speed internet can support real-time information sharing and modifications of big data sets, such as digital building models, which are often required in large-scale projects. Remote design collaboration can significantly reduce operational costs and help increase efficiency in global design teams. Early development of these systems including, for example, DesignWorld (Maher et al., 2006), demonstrates the concept of a collaborative platform for synchronous communication, collaborative modelling and multidisciplinary information sharing.

4.3.2. Experiments and coding scheme

The three different design environments developed and simulated in the experiments in this case study include a traditionally co-located sketching environment using pen and paper, a remote sketching environment using Smart Board and Mimio and a remote modelling environment using Active Worlds. In the co-located sketching session, the pairs of designers used pen and paper as the tools. In the remote sketching session, they could sketch synchronously from remote locations, with one designer using pens on a vertically mounted Smart Board (www.smarttech.com) and the other using a pen interface on a table-top system using Mimio hardware and software (www.mimio.com). Both designers were provided with a pen and digital ink interface to a 2D surface with a projection of the design model. In the final 3D virtual world session, designers collaborated synchronously from remote physical locations but with a shared virtual presence in a 3D virtual world – Active Worlds (www.activeworlds.com) – through 3D design and modelling. For each of the three experimental environments, the participating pairs of designers were provided with a design brief and a collage of the photos showing a site and its surroundings. In the latter two sessions of the experiment, remote communication was simulated by locating the participants in different parts of the same room, in such a way that they could hear each other, but only see each other via web cams. This was done to reduce bandwidth problems, so that the focus could be on cognitive activities rather than technological limitations.

Each session required the designers to complete a simple architectural design task in 30 minutes. The designers were also given training in the use of Smart Board, Mimio and ActiveWorlds tools prior to the experiment. A five-category coding scheme – including communication content, design process, operations on external representations, function–structure, and working modes – was used for the study and this is described in the following points. (i) The communication content category partitions each session according to the content of the designers' conversations, focusing on the differences in the amount of conversation devoted to discussing design development when compared to other topics. (ii) The design process category characterises the different types of designing activities that dominate in the three different design environments. (iii) The operations on the external representation category look specifically at how the designers interact with the external design representation to see if the use of 2D sketches or 3D models is significantly different. (iv) The function–structure category further classifies the design-related content as a reference to the function or structure of the design. (v) The working modes category characterises each segment, according to whether or not the designers are working on the same design task, or on the same part of the design representation.

4.3.3. Protocol analysis results and discussion

The analysis of collaborative design behaviour involves documenting and comparing the categories of codes using INTERACT software. First, the frequencies of the occurrence of the code categories in the three different sessions are examined. Next, the time spent for each category is determined, with respect to the total time elapsed during the session. This data gives us the duration percentages of the codes in each main category. To examine the validity of the results, a test of significant differences is undertaken between the pairs across the three design sessions in terms of the design behaviour (coded activity categories). The ANOVA test result (ANOVA with replication, $p<0.05$ between the three different design sessions) shows that there is no significant difference between the pairs in terms of communication content ($p=0.15$), operations related to external representations ($p=0.80$) and working mode ($p=0.99$). Thus, across the different pairs, the collaborative behaviour in these categories is similar. In contrast, the design process (2.46E-08) and function–structure (3.26E-05) categories are significantly different between the pairs. Such differences are common in design research where the activities of a particular designer can evolve in response to a particular context, which can have an effect on the collaborative design process.

Table 4.5 provides an overview of the focus of activity in each of the three design environments, by showing the average percentages for four of the five coding categories, of the pairs of the designers who participated in the experiment. The working mode category is not included because it is always 100% of the duration, since the designers are always either working on the same or on different tasks. The categories of codes were applied independently, therefore each segment could be coded in more than one category.

Table 4.5. Durations of codes in each main category as average
percentages of the total elapsed time.

Categories	Face to face sketching	Remote sketching	3D virtual worlds
Communication content	72%	72%	61%
Design process	69%	48%	34%
Operations on external representations	96%	90%	93%
Function–structure	67%	43%	27%

The data reveals there are significant differences across the three design environments in terms of function-structure (40%) and the design process (35%), and a smaller range of differences in communication content (11%) and operations on external representations (6%). The result for communication content is potentially important as it indicates that design communication and representation during collaboration are possible, even when designers are remotely located. In terms of the design process and function–structure categories however, there is a significant decrease from co-located sketching to remote sketching and to remote 3D design and modelling in virtual worlds. While on the one hand, this result confirms that design collaboration can occur with appropriate technical support in remote environments, on the other hand, the results suggest that their experience and process has some significant differences. For example, the results indicate that during the co-located session, designers mostly worked closely together, with over 95% of the session devoted to collaborative activities. Although sometimes a particular designer led the process while the other observed and critiqued, they always focused on the same tasks. In the remote sessions, and especially in 3D virtual worlds, an average 40% of the duration was for individual design activities where different designers worked on separate tasks or different parts of the design representations. The pairs would then come together to review each other's outcomes or exchange tasks. During these individual phases, they either reduced or even stopped verbal communications, which explains the decrease in communication content in the 3D virtual environment. This also explains the reduced number of design-related segments in both the remote sketching and 3D virtual environments. The decrease in the design process and function–structure categories from co-located sketching to remote sketching and to remote 3D virtual worlds may also reflect the differences between 2D and 3D thinking and representation. In both sketching sessions, while designers constantly externalised their ideas in separate sketches, or as additions over existing sketches, there was usually a clear separation between the development of concepts and of formal representations. In 3D virtual worlds, however, these processes overlap, as designers re-used their conceptual models as the basis for design documentation. In this particular environment the design ideas are conceptualised, explored and refined using the same 3D modelling.

Returning to the initial question asked in this study: does the nature of the collaborative environment have an impact on the behavioural and cognitive operations of a design team? In terms of communication, ironically, no, it does not have a quantifiable impact, although there may be quality issues that are not identified in this study. However, in terms of the design process and function-structure collaborative behaviours, there is a substantial difference. Possibly the most obvious difference is that in the two sketching environments, drawings were cyclically produced, modified and discarded after interrogation or testing. In the virtual world environment, the concept model effectively evolved into the schematic model through a series of cycles, each one directly building on the last. These two processes, despite what at first glance might seem to be only minor differences (a process of cyclical sketching is replaced with a process of cyclical modelling), actually involve different types of communication, representation and cognition.

4.4. Case study 3: A biometric approach to analysing cognitive behaviour in a CAD environment

The early modernist architect Richard Neutra famously argued that vision precedes action, or that our eyes lead and our bodies follow (Ostwald and Dawes, 2018). This tendency has recently begun to be studied using eye-tracking technology, demonstrating that our eyes seek out a target or option before we move or respond (Jacob and Karn, 2003). Importantly, data derived from eye-tracking is thought to offer a clear indicator of cognitive behaviour and intent (Ware and Mikaelian, 1987). Marr (Marr 1980, 2010) laid the foundations for the study of cognitive visual processing, using edge detection, image intensity changes and object recognition. In the design field, past research using eye-tracking has focused on how designers look at a static scene (Kaufman and Richard, 1969; Weber et al., 2002; Yu and Gero, 2017).

While there is a growing body of knowledge about the ways people view the environment and respond to it, there is a lack of empirical evidence regarding designers' vision and response activities during the design process. There is also a particular gap in our knowledge about the way designers use CAD, both in terms of specific tools and as a reflection of the cognitive processes and behaviours that occur in the

CAD environment. To address this knowledge gap, this exploratory study focused on the relationship between eye movement and tool use in computer-aided design. The data is developed from a pilot experiment where two participants completed an architectural design task in a CAD environment. The design activities were captured using protocol analysis and eyetracking.

4.4.1. Experiment

In this study, two masters level architecture students from Harbin Institute of Technology, China, were recruited as participants. Senior architecture students were selected because they have the architectural design experience and skills to execute a complete design process. In the experiment, they were required to complete a defined architectural design task using computational modelling software, which in this experiment was *Sketchup*. The participants each had more than five years' experience using *Sketchup*.

During the experiment, the designers' activities and their verbalisations were video-recorded and the data subsequently used for protocol analysis. The designers were given 60 minutes for the design session. The design task was a conceptual design for a high-rise building, with specific functions listed and located on a pre-modelled site that was provided to the participants. During the experiment they were not allowed to sketch manually and, thus almost all their actions occurred on the computer, thereby ensuring that the design environment could be captured for analysis. The participants' eye movements were recorded using Tobii Studio eye-tracking equipment. Figure 4.20 shows the experiment set up.

Figure 4.20. Experiment setting (Photos were taken by authors).

4.4.2. Results

This protocol study employed an integrated segmentation and coding method. The segmentation and coding processes are based on the "one segment one code" principle (Kan and Gero, 2017). This means there are no overlapped codes or multiple codes for any segment. If there are multiple codes for one segment, the segment is further divided.

Two sets of coding schemes were applied in this research, the first is based on the designer's verbalisation/activities, which is founded on the FBS ontology. The second coding scheme is based on the designer's eye-tracking data, which gives the location of their gaze. Table 4.6 shows the results of the segmentation.

Table 4.6. General results.

	Total segments (FBS)	Total segments (Eye-tracking)	Design time (Mins)
Designer 1	360	1191	49.03
Designer 2	357	1118	36.29

In order to obtain a more detailed articulation of what the designer's focus was on when modelling, *Structure (S)* is divided into three sub-categories: Ss (site), Sb (base) and St (tower). The design issue distribution for each designer, derived by applying the FBS-based coding scheme, is presented in Table 4.7. The data in Table 4.7 indicates that both designers expended most of their design effort on *structure (S)* related activities. Within the structure category, both designers expended more cognitive effort on St (53.1% and 23.0%), which is the geometry of the tower, followed by Sb (base, 14.2% and 20.2%) and Ss (site, 9.4% and 4.8%). After *structure (S)*, the second highest effort was expended on *Bs* (13.1% and 24.7%), which is related to the examination of the design as it is generated.

Table 4.7. Design issue analysis.

	R (%)	F (%)	Be (%)	Bs (%)	Ss (%)	Sb (%)	St (%)
Designer 1	1.9	3.6	4.7	13.1	9.4	14.2	53.1
Designer 2	1.1	7.0	19.1	24.7	4.8	20.2	23.0
Mean	1.5	5.3	11.9	18.9	7.1	17.2	38.0

The coding scheme for the eye-tracking data was developed based on an examination of the eye gaze data. It used the location of the eye gazes in the design, Table 4.8.

Cumulative analysis of protocol and eye-tracking data

The cumulative occurrence of design issues represented by the codes is determined from the segmentation and coding of the design session (Gero and Kannengiesser, 2012). The cumulative occurrence graph provides a basis for the qualitative evaluation of the expenditure of design effort, which correlates to the cognitive effort, during a design session (see Figures 4.21 and 4.22). A visual inspection of the results in these figures shows that for Designer 1 the predominant activity was associated with the tower (St),

Table 4.8. Coding scheme for eye-tracking data.

		Code
Site planning	Surrounding of the site	Ss
	Corner	Sc
	Edge	Se
	Face	Sf
Base	Corner	Bc
	Edge	Be
	Face	Bf
Tower	Corner	Tc
	Edge	Te
	Face	Tf
Menu		M

which was iteratively developed throughout the entire session. In parallel, this designer worked on the design of the base (Sb) for the first 30% of the session only (i.e. up to decile 3). Designer 2 also initially worked on the base (Sb) for 50% of the session, but only commenced designing the tower (St) from this point. This shows the different sequence of designing.

For both designers, St and Bs activities increased towards the end of the design session, which means that the designers committed increased cognitive effort to the geometry of the tower until the end of the design session. Conversely, they worked on the base and site of the building more at the beginning than the end. For both designers, F increased more rapidly at the beginning of the session. This suggests that the designers tended to consider the function-related issues at the start of the process, which matches the results of previous cognitive research on design (Gero and Kannengiesser, 2012; Yu et al., 2013).

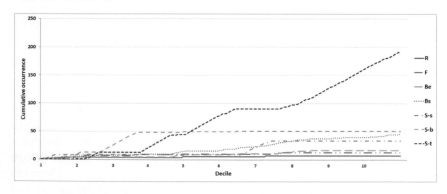

Figure 4.21. Cumulative analysis of FBS coding (Designer 1).

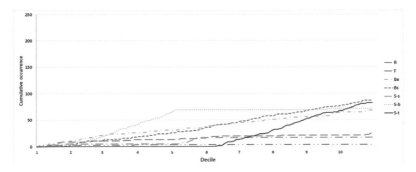

Figure 4.22. Cumulative analysis of FBS coding (Designer 2).

The cumulative analysis of eye-tracking data is produced by segmenting and coding the way the eye gazes on façades, edges and corners of the building (Figures 4.23 and 4.24). From these results, both designers focused their gaze more on the façades (723 gazes and 733 respectively), followed by edges (245 and 241) and then corners (144 and 55). In architectural design, shape and space are two fundamental primitives usually considered by designers (Ching, 2014; Moore and Allen, 1977). Previous research shows that designers' eye movements tend to focus more on the edges of static images than the spaces they define (Weber et al., 2002). Results from the present study of a dynamic design process suggest the opposite. Thus, when Designer 1 was working, her attention was more on the spaces than the edges, and only rarely on corners.

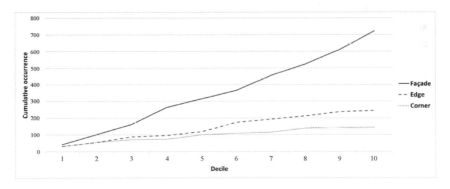

Figure 4.23. Cumulative analysis of the eye-tracking gazes on façades, edges and corners (Designer 1).

The cumulative analysis of eye-tracking data derived from the two designers (Figures 4.25 and 4.26) shows that both exhibited increases in Tower façade (Tf) gazes towards the end of the design session (up to

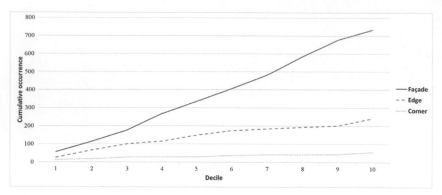

Figure 4.24. Cumulative analysis of the eye-tracking gazes on façades, edges and corners (Designer 2).

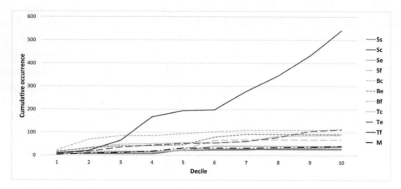

. **Figure 4.25.** Cumulative analysis of eye-tracking coding (Designer 1).

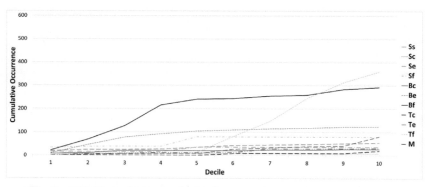

Figure 4.26. Cumulative analysis of eye-tracking coding (Designer 2).

541 and 360). This means that they worked on the façade of the Tower throughout the session, with an increase of attention towards the end of it. For Designer 1, visual attention to Tf dominates the entire session, and for Designer 2, Tf dominates the second half of the session. Visual

attention to the Base façade (Bf) is the next most dominant type, whereas the designers' gazes were only rarely focused on the screen menus.

Heatmaps of eye movement

Heatmaps generated for eye movement (Figures 4.27 and 4.28) demonstrate that most of the gaze directions are to the middle of the screen, some of the locations are towards the left-hand side and only rarely does each person's gaze focus on the right-hand side. This confirms Arnheim's

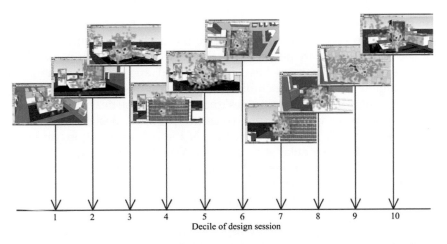

Figure 4.27. Designer 1, heatmaps of the eyetracking during design in deciles. Red-yellow-green represents the gradation of time focused in that region, with red the highest, yellow the middle and green the lowest.

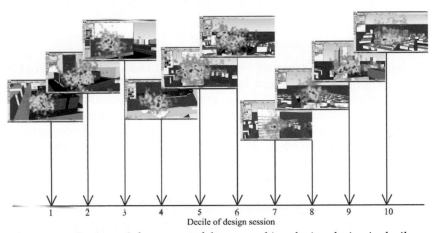

Figure 4.28. Designer 2, heatmaps of the eyetracking during design in deciles. Red-yellow-green represents the gradation of time focused in that region, with red the highest, yellow the middle and green the lowest.

(1974) and Weber et al.'s (2002) results that participants focus usually on the left-handside of a picture. Another finding is that most of the tracked gazes are focused on the model, and only a few on the menu; this may be because the participant generally used shortcuts in *Sketchup* rather than menu commands.

4.5. Case study 4: Implementing rules in design, using generative design grammars

By rigorously analysing existing designs of a known style, design grammars can be developed to describe the design style and these can also be used to generate new designs that share the style. For nearly four decades, design grammars have been refined and examined across a wide range of disciplines, including architectural design, interior design, product design and engineering design. Grammars are design formalisms that sequentially apply simple rules to produce designs with rich properties. By alternating the sequence of the rule application, different designs that share a similar style emerge. With the increasing availability and accessibility of computer technologies, design grammar research has evolved from being a largely manual implementation of mathematical models to being a practical computational process for generating and analysing design.

This case study presents an approach to automating and optimising the generative design process through an integrated Generative Design Grammar (GDG) framework supported by agent reasoning and design simulation. This approach to grammatical thinking in design offers both opportunities and challenges for designers working in this environment or context.

4.5.1. Design grammars

The concept of GDGs adopted for the present research was inspired by shape grammars as a design formalism for describing and generating designs (Stiny and Gips, 1972). Shape grammars, which are both descriptive and generative in nature, have two components. The first are the shapes (points, lines, planes or volumes) and the second are the rule descriptions that modify shapes to create designs. The applications of the shape rules generate designs by way of shape operations and spatial transformations. In combination, these features are core to GDGs. In architectural design there are many successful shape grammar applications; for example, the Palladian grammar (Stiny and Mitchell, 1978), the Mughul Gardens grammar (Stiny and Mitchell, 1980), Prairie Houses grammar (Koning and Elzenberg, 1981) and the Siza Houses grammar (Duarte, 1999). Knight (2000) defines a shape grammar as a set of shape rules that can be applied in a step-by-step manner to generate a set, or language, of designs.

The GDG described in the present case study is capable of describing designs in 3D virtual environments using components of the design rules and generating designs via rule applications. The descriptive and generative qualities of the GDG support designing in 3D virtual environments, and also have applications in any design environment where rules are used to generate solutions.

4.5.2. The conceptual framework of generative design grammars

This section presents a framework that provides guidelines and strategies for developing a GDG by defining a generic structure for a grammar and its basic components and rules. Using this framework, designers define grammars that produce different design languages. In this case study they are demonstrated in 3D virtual environments, where it may be preferable to predefine rules for automated generation rather than defining the detail of all possible designs.

A GDG is comprised of design rules R, an initial design Di, and a final state of the design Df.

$$GDG = \{R, Di, Df\}$$

The general structure of a GDG comprises four sets of design rules: layout rules Ra, object placement rules Rb, navigation rules Rc, and interaction rules Rd.

$$R = \{Ra, Rb, Rc, Rd\}$$

The structure of a GDG is determined by the four general phases of designing.

(1) *To layout places/areas of the design*, such that each place/area has a purpose that accommodates certain intended activities.
(2) *To configure the places/areas of the design*, such that each place/area is configured with objects that provide the visual boundaries of the place/area and visual cues for supporting any intended activities.
(3) *To specify navigation methods* such that navigation in the design can be facilitated by providing wayfinding aids for supporting designers and visitors to explore and understand the design.
(4) *To establish interactions* by ascribing behaviours to selected objects in each place/area of the design such that physical interactions of the design can be simulated or the designers and visitors can interact with the virtual design.

These four sets of design rules – layout, object placement, navigation and interaction rules – may be mapped to four design phases in the GDG framework (Figure 4.29).

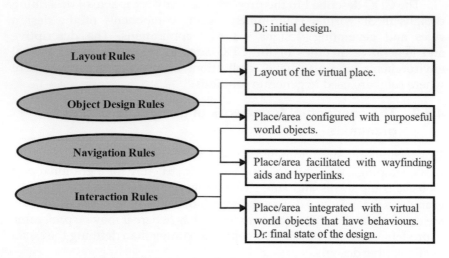

Figure 4.29. Generative design grammar framework.

The sequence of the design rules commences with the order of layout, then object placement, navigation and finally interaction rules. When integrated with relevant design and domain knowledge, the GDG can be developed by following this general structure. The stylistic characterisations of the generated designs – in terms of the syntax (visualisation, layout and object placement) and in terms of the semantics (navigation and interaction) – are defined accordingly in these four sets of rules.

4.5.3. Design rules

The first component of the GDG is the set of design rules. The general structure of design rules is similar to the general structure of shape rules. In shape grammars, a shape rule can be defined as:

$$LHS \rightarrow RHS$$

which specifies that when a left-handside shape (LHS) is found in the design, it will be replaced by a right-hand-side shape (RHS). The replacement of shapes is usually applied under a set of shape operations or spatial transformations. The shapes are labelled (the use of spatial labels and state labels) for controlling the shape rule applications. Similarly, a design rule of GDG is defined as:

$$LHO + sL \rightarrow RHO$$

which specifies that when a left-hand-side object (LHO) is found in the 3D virtual environment, and the state labels sL are matched, the LHO will be replaced by a right-hand-side object (RHO). The term "object" can refer to a virtual object, a set of virtual objects or virtual object properties. Virtual objects are visualised as 3D models in 3D virtual environments. Like shape grammars, GDGs also use spatial labels and state labels to control the application of design rules. The original use of state labels in a shape grammar is to control the sequence of shape rule applications. In a GDG this original purpose is maintained such that the design's rules can be applied in the sequence of layout rules, object placement rules, navigation rules and interaction rules. In addition, in the GDG a special set of state labels are also developed, as discussed hereafter. The general structure of design rules has the following two considerations.

(1) State labels are singled out and expressed explicitly as sL in the structure. The use of state labels is critical to the GDG, as through their application they direct the grammar to ensure that the generated design satisfies the current design goals. Each design rule is associated with certain state labels representing specific design contexts that can relate to different design goals. In order for a rule to be applied, a virtual object, a set of virtual objects or a virtual object's properties need to be found in the environment that match the LHO of the design rule, and the design contexts represented by the sL of the design rule need to be related to the current design goals.

(2) The basic components of design rules are objects and their properties, not shapes. Therefore, they are not always visual or spatial. For the interaction rules and parts of the navigation rules, the replacement of LHO with RHO is applied under a set of general transformations.

Layout rules are the first set of design rules to be applied in the GDG framework. They are visual or spatial rules that generate the arrangement of places/areas according to the activities they are intended to support, as defined in the design requirements. Figure 4.30 illustrates two example layout rules for a GDG to support the design of a gallery. By applying the first rule, the design of the gallery is expanded to add an additional area. By applying the second rule, the design of the gallery is changed by subtracting an area.

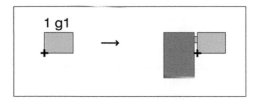 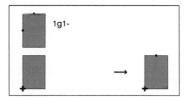

Figure 4.30. Two examples of layout rules.

Object placement rules are applied after layout rules and they, too, are fundamentally visual or spatial in nature. After a layout is produced, object placement rules further configure each place/area to define the visual boundaries of the place/area and visual cues, through object placements, for supporting intended activities. Figure 4.31 shows two example object placement rules that generate the visual boundaries for two different areas in a gallery design. Figure 4.32 shows an example object placement rule that arranges the interior of a display area for an exhibition.

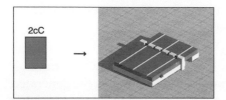 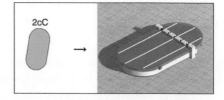

Figure 4.31. Two example layout rules that generate visual boundaries for different areas of a gallery design.

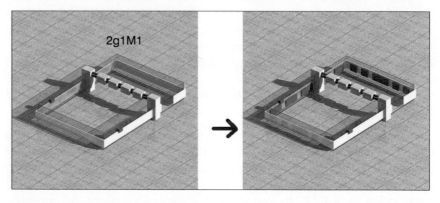

Figure 4.32. An example layout rule that arranges the interior of a display area for exhibition.

Navigation rules are the next ones applied in the GDG. They provide wayfinding aids to assist designers and visitors to navigate around the generated places/areas. Wayfinding aids can be simulated in virtual environments by drawing on experience in the built environment (Darken and Sibert, 1993, 1996; Vinson, 1999). There are at least two kinds of wayfinding aids commonly used in built environments. The first are spatial cues, for example paths, openings, hallways, stairs, intersections, landmarks, maps and signs. The second are social cues, including, for example, assistance provided by guides or other people. In addition to these conventional ways of navigating built environments, virtual environments also have their own unique tools to assist navigation. For example, most 3D virtual environments allow people to move directly between any two locations using hyperlinks. Hyperlinks are not part of a design, as they cannot be reproduced in real built environments. They are, however, common in 3D virtual environments as they often enable designers and visitors to explore more efficiently. As this last factor suggests, navigation rules are not necessarily just about visual or spatial properties. Navigation can be supported by sound or even music in virtual words, and in the real world smell can play a major role in wayfinding.

Navigation rules are typically concerned with recognising the connections between places/areas and the ways designers and visitors can access and cognitively appreciate the spatial structure of these. Figure 4.33 shows the effect of an example navigation rule. The left-hand-side is the interior of a display area in a gallery design. The right-hand-side shows that a hyperlink is created and appears as a coloured stone on the floor. After appropriate behaviours are applied, the link will take the designers and visitors who deliberately step on the stone to a different display area.

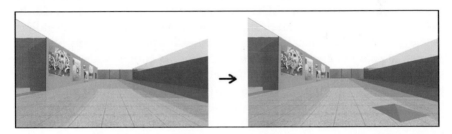

Figure 4.33. The effect of an example navigation rule.

Interaction rules are the final set of design rules applied in a GDG. They ascribe behaviours to selected objects such that physical interactions can be simulated or the designers and visitors can interact with the design. There are two different types of interaction rules. The first supplements object placement rules and the second supplements navigation rules. The first ascribes behaviours to relevant objects in order to simulate certain interactions. The second supports wayfinding and hyperlinks generated by navigation rules and ascribes appropriate behaviours to activate them.

Interaction rules are innately non-visual and spatial; rather, they are behavioural or responsive, even though some responses might be visually apparent. For example, Figure 4.34 shows the effect of an interaction rule of the first type for supplementing object placement rules. The left-hand-side image is the exterior of a gallery design with an empty advertisement board. The right-hand-side image shows the same advertisement board displaying digital images in an animated sequence after the interaction rule is enacted. This configures the object properties of the advertisement board using a scripting language to enable the animation to be shown.

Figure 4.34. The effect of an example interaction rule.

4.5.4. Designing a virtual gallery

This section presents an example GDG scenario for designing a gallery. The scenario aims to demonstrate dynamic and autonomous design in a 3D virtual environment. GDG, as a generative design system, can be manually applied by human designers, or as demonstrated in this gallery design scenario, dynamically and automatically applied by computational agents. In computer science, agents as embodied components of software systems operate independently and rationally, seeking to achieve goals by interacting with their environment (Wooldridge and Jennings, 2020). Unlike most computational objects, agents are programmed to have "goals" and "beliefs" and execute actions based on these (Russell and Norig, 1995). The agents used in this demonstration are called Generative Design Agents (GDA) (Gu and Maher, 2005). GDAs are rational agents enabled by five computational processes: sensation, interpretation, hypothesising, designing and action. These processes provide a basis for allowing design and other domain knowledge to be integrated into GDA, which together support reasoning and designing.

The design scenario demonstrated in the final stage of this case study consists of six different stages. The stages present various changes to design requirements during the designing process, for example, changes of activities, exhibition requirements, gallery capacities and so on. The scenario shows that a GDA can respond to these changes, either by simulating alternatives in the 3D virtual environment or by being entered directly by a human designer. By altering the sequence of the design rule application in the GDG, the GDA dynamically and autonomously generates, simulates and modifies different designs in the virtual environment.

The technical implementation of this case study used *Java* and the scenario was situated in a 3D virtual environment developed using the

Active Worlds platform. The design rules of the GDG example, and a general rule base for supporting the GDA's reasoning, were written using *Jess* (jess.sandia.gov), a rule-based scripting language (Friedman-Hill, 2003). Figure 4.35 illustrates different designs for the gallery, both 3D models and plans, generated by the GDA through the application of the GDG over six stages of the scenario.

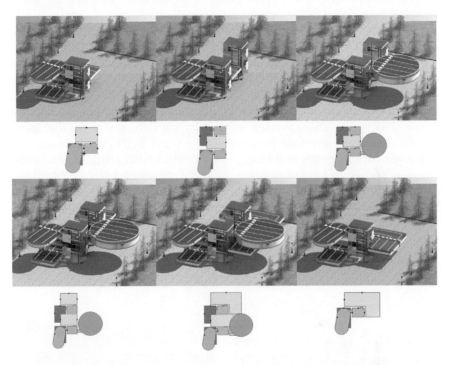

Figure 4.35. 3D models and plans of a virtual gallery generated for stages 1–3 (above) and stages 4–6 (below) of the design scenario.

If there is more than one design rule that matches the current design context, a control mechanism is required to resolve the conflict. In general, there are three main methods for controlling the application of the GDG. The first of these is *random selection*, which programs the system to randomly select one design rule from the set of available rules that meet the conditions. The second is *human intervention*, wherein the designer provides instructions if a conflict occurs. This case study scenario uses the human designer intervention method. The third control uses *agent learning* to allow the system to resolve the conflict based on the agent's past design experience.

4.6. Case study 5.1: Generating new design instances of an existing style using computational analysis

The penultimate case study in this chapter is an example of an analytical and generative approach, wherein the design environment is a combination of three computational design methods: syntactical analysis, fractal analysis and parametric design. The focus of this case is the traditional Chinese private garden (TCPG) (Hunt, 2012; Rinaldi, 2011).

The origin of the TCPG is typically traced to the Tang Dynasty (circa 800 AD), although examples of private hunting reserves with artificially constructed landscapes can be found in China in the 11th century BC. By the 17th century AD, the TCPG had become a special type of landscaped space, characterised in part by its rich spatial arrangement and its associated aesthetic properties. A typical TCPG is made up of a dense network of paths and spaces, punctuated with artificial landscape features, ponds and small streams, paved squares and covered bridges, all of which are organised in a relatively small and clearly defined area (Figure 4.36). Today the TCPG is renowned for exhibiting high levels of spatial complexity and variety, properties that have led them to being accepted as having unique experiential qualities (Peng, 1986; Tong, 1997).

Figure 4.36. Yuyuan Gardens in Shanghai (photos were taken by authors).

Designers and researchers have analysed the TCPG from various qualitative perspectives (Hunt, 2012; Tong, 1997) and a small number of studies have also examined their spatial properties using quantitative methodologies (Chang, 2006; Li, 2011; Lu, 2009, 2010; Wang and Wang, 2013). For example, in the former case, Keswick (1978) and Zhou (1999) separately explored the spatial character of TCPGs from a historical and social perspective. In the latter case, using quantitative methods, multiple attempts have been made to understand the dense spatial configurations in TCPGs and the changing vistas experienced while navigating through them (Chen, 2012; Li, 2011; Peng, 1986). Syntactical techniques, which analyse the topology of connectivity of space, have also been employed to examine TCPGs (Chang, 2006; Chen, 2012; Wang and Wang, 2013) and to support grammatical interpretations of their formal compositions (Lu, 2009, 2010) or visual perceptual characters (Li, 2011).

Despite this growing body of research, the spatial qualities that make TCPGs unique have rarely been measured and generalised mathematically. Without such a set of measures, their spatial and aesthetic properties cannot be replicated in current landscape design practice or maintained as part of the restoration process for these sensitive heritage structures. Thus, the purpose of the present case study is to begin to develop a system that can generate new TCPG plans using parametric design and which capture selected aspects of the spatial structure and visual character of the originals. The initial focus of this case study is a set of three 16th century TCPGs in southern China: Yuyuan (in Shanghai), Zhuozhengyuan (in Suzhou) and Liuyuan (in Suzhou). These three are all typical examples of Ming and Qing dynasty private gardens (Fu et al., 2002) and have total areas of over 20,000 m².

This case is structured in three stages that replicate the steps in the research method. The first stage, syntactical derivation, uses connectivity graphs to measure three properties of each TCPG. As part of the process, different garden spaces are categorised into various spatial types – effectively, different kinds of areas in the landscape. An inequality genotype is then developed for each TCPG and the mathematical value ranges of the measurements for each spatial type are identified. These values are used in the second stage, parametric generation, where rules are developed to create nine new designs that feature the same spatial structures found in the historic TCPGs. In the final stage of the research, the new designs and the three originals are measured using the box-counting method to determine their fractal dimensions. This allows for the characteristic or typical spread of visual information in each to be compared, providing a mathematical assessment of the degree to which the new cases visually resemble the historic ones. Each of the three methods used in the chapter – connectivity graphs, parametric rules and the box-counting method – are presented at the start of their respective stages. The ultimate purpose of the three-stage process is to begin to develop and test a method for generating new designs that conform to the styles and spatio-structural qualities of an existing body of work.

4.6.1. Stage 1: Syntactical derivation

Space syntax is a theory and associated set of techniques that developed unique architectural and urban applications of graph mathematics for the analysis of the social structure of space (Hillier, 1995; Hillier and

Hanson, 1984). Extensively refined over the last few decades, space syntax methods have been widely applied in urban planning research, architectural design and landscape design, amongst other areas. One of the key characteristics of the space syntax approach is that it provides a way of understanding architectural and urban spatial configurations by translating their properties into topological graphs that can then be mathematically analysed and socially interpreted (Hanson, 1998).

One of the earliest syntactical analytical techniques, known as a Convex Graph or a Justified Plan Graph, creates a graph from a set of nodes which represent spaces, and the connections between them, representing a type of boundary condition that might be trafficable, permeable or visual depending on the chosen application. There are multiple variations of this technique, including those pertaining to visually-defined spaces ("convex spaces"), functionally defined rooms, or functional zones in a plan or a landscape (Bafna, 2003; Hanson, 1998; Minor and Urban, 2007). In all cases, once the chosen spaces are represented by a graph, various mathematical properties of the graph may then be derived from it (Hillier and Hanson, 1984; Hillier and Kali, 2006; Ostwald, 2011). For example, the step depth of each node in the graph can be determined, then the total depth (TD), mean depth (MD) and integration over values calculated for each node. These values reflect the relative role of each space within the larger structure of the plan. For example, the step depth suggests the connectivity distance from the entrance node, the mean depth is the average depth for each node, which represents the degree of isolation of the spaces, while the degree of integration is suitable for comparing each space with the other parts in a distributed plan, and can be used to develop an inequality genotype, a hierarchical determination of the structure of the spaces in a graph (Bafna, 2001).

While space syntax techniques have typically been formulated to analyse spaces that have underlying functional properties associated with efficiency or control (of movement or access), the TCPG is not structured in this way. Arguably, the purpose of the TCPG is to provide a space for informal wandering and contemplation. The design of the TCPG emphasises the joy of an uncertain spatial experience, with paths snaking between areas in the garden seemingly serving to evoke a sense of delight or mystery (Keswick et al., 2003). To do this, the TCPG features several recurring design elements and strategies that collectively constitute a spatial pattern. For example, the pavilions and covered spaces in the TCPG are often strategically placed around bodies of water, which are themselves sited at the centre of the garden. Within the trafficable parts of the gardens, large courtyards are commonly surrounded by smaller spaces (Guo, 2014). Covered or semi-enclosed structures punctuate the networks of paths between these courtyards, including pavilions lined with lattice windows and covered narrow corridors (Sun, 2012). These spatial configurations and relations, which are repeated in variations in many TCPGs, provide the unique experiential quality observed by people when moving through the gardens (Keswick et al., 2003). They also evoke the sense of a linear navigation path that is sequenced or organised to create a changing rhythm that shifts from small, tight spaces, to larger, open ones and then back again (Chen, 2012; Guo, 2014). Because of these properties of the TCPG, for the present case a variation of the standard connectivity graph technique for analysing functional areas was selected. However, because TCPGs do not possess distinct functional zones – effectively all spaces serve as passive recreation or contemplation areas – their areas were differentiated in terms of spatial types in accordance with conventional readings of the features (Guo, 2014; Keswick et al., 2003).

For the present case, each of the three historic TCPGs are analysed and a set of six recurring spatial types identified: (1) large rooms, (2) small rooms, (3) pavilions, (4) yards and squares, (5) covered corridors and (6) pathways. Large and small "rooms" are physically and often visually defined in plans by walls and landscape elements. In the three TCPGs studied, there were two clearly distinguishable types to these rooms, hence the "large" and "small" descriptors. However, the definition of a large or small room is not solely dependent on its area. Both area and inclusions or constituent parts are considered as part of the classification process. Thus, the larger room is indeed typically bigger in area than the smaller one, but importantly, it also usually features a courtyard with some landscape elements in it. Therefore, large rooms are those which provide a relatively complete and independent living space, whereas small rooms, even if they are close to the area of their bigger counterparts, do not have these features. The spatial type "pavilion" includes covered or roofed structures (typically without extensive walls) which dominate a defined "room" in the garden. "Yards and squares" are open paved spaces which are less clearly defined, and are smaller in scale than the garden "rooms". The final two types are variations of long narrow spaces, the first of which is a roofed or "covered corridor" and the second is an uncovered, often meandering "pathway". The extent of these corridor types is defined by the surface treatment of the ground plane (typically paving stones) and low landscaped walls and planting.

With these six spatial types as a basis, plan analysis of the three historic TCPGs was undertaken using *UCL Depthmap* software. Figure 4.37 shows the Yuyuan garden, which is colour coded and annotated to identify the six spatial types. In Figure 4.38 the exterior is presented as the carrier level in the graph, and the principle for calculating the total depth (TD) of a space requires that, if it can fit into different depth levels, the priority is given to the lowest level. The mean results were derived for each measure for each spatial type and the standard deviation recorded (Tables 4.9-4.11). Each of the three sets of results are described hereafter.

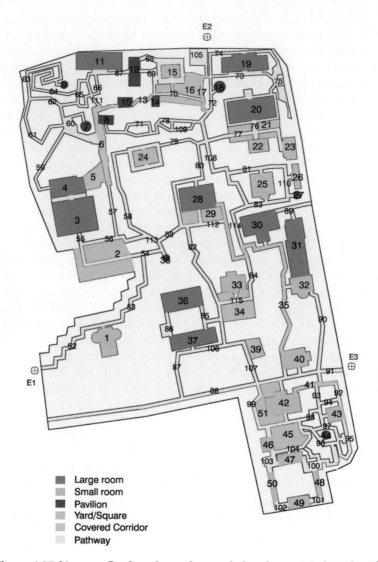

Figure 4.37. Yuyuan Garden plan, colour coded and annotated to identify the six spatial types.

In the Yuyuan garden results (Table 4.9) the pavilion has the highest *MD* value (*MD* = 4.73), which means it is the most isolated spatial type, and the covered corridor has the lowest *MD* value (*MD* = 3.25). The integration value suggests that the pathway is the most integrated space (*i* = 12.04), and the least is the covered corridor (*i* = 0.67). The results confirm the common expectation because in TCPGs the pathway generally provides the major connection to other spatial types.

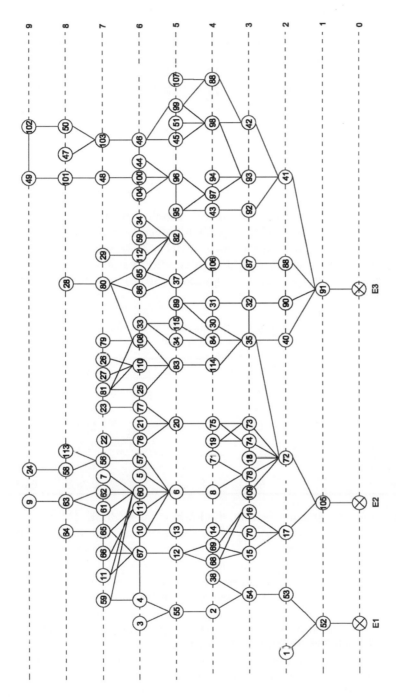

Figure 4.38. Yuyuan Garden, connectivity graph representation of the data.

Table 4.9. Connectivity analysis of Yuyuan Garden (mean results for *TD, MD* and *i* for each spatial type).

Spatial type	Number of spaces	TD	MD	SD of MD	i
Large room	10	39	4.33	1.29	1.20
Small room	18	79	4.65	2.28	2.19
Pavilion	12	52	4.73	1.56	1.34
Yard/Square	7	22	3.67	2.54	0.94
Covered corridor	5	13	3.25	1.67	0.67
Pathway	63	219	3.53	2.31	12.04

In the Zhuozhengyuan garden results (Table 4.10) the pavilion type has the highest mean depth (MD = 6.17), followed by the courtyard (MD = 6.10) and the path type has the lowest (MD = 3.98), which is the same pattern as in the Yuyuan garden. The integration value also shows that the pathway is the most integrated space (i = 10.05) and the least is the covered corridor (i = 0.11).

Table 4.10. Connectivity analysis of Zhuozhengyuan Garden (mean results for *TD, MD* and *i* for each spatial type).

Spatial type	Number of spaces	TD	MD	SD of MD	i
Large room	9	42	5.25	2.45	0.82
Small room	14	64	4.92	2.06	1.53
Pavilion	13	74	6.17	1.93	1.06
Yard/Square	11	61	6.10	2.02	0.88
Covered corridor	3	11	5.50	1.15	0.11
Pathway	62	243	3.98	2.38	10.05

In the Liuyuan garden results, the highest mean depth (MD = 8.40) is for the covered corridor, followed by the pavilion (MD = 6.80) (Table 4.11). This differs from the other TCPGs tested, because in the Liuyuan garden the covered corridor is the most isolated space. However, as with the other two, the integration values suggest that the pathway is the most integrated space (i = 3.22) and the covered corridor is the least (i = 0.27).

The results for the three TCPGs suggest the following trends. First, for all of the gardens the most integrated spatial type is the pathway, followed by the small garden room, and the least integrated is the covered corridor. Second, for the Yuyuan and the Zhuozhengyuan gardens, the pavilion type has the highest mean depth, and for the Liuyuan garden the highest mean depth is the covered corridor, followed by the pavilion. Finally, for

Table 4.11. Connectivity analysis of Liuyuan Garden (mean results for *TD, MD* and *i* for each spatial type).

Spatial type	Number of spaces	TD	MD	SD of MD	i
Large room	7	36	6.00	1.35	0.50
Small room	13	81	6.75	2.28	0.96
Pavilion	6	34	6.80	2.34	0.34
Yard/Square	10	52	5.78	2.66	0.84
Covered corridor	6	42	8.40	0.89	0.27
Pathway	33	186	5.81	2.38	3.22

two of the gardens (Zhuozhengyuan and Liuyuan), the smallest mean depth value is the pathway. For the Yuyuan garden the smallest mean depth value is for the covered corridor. These results confirm that for the different spatial types in these TCPGs, there are multiple common mathematical characteristics that can be generalised.

TCPG inequality genotype

An "inequality genotype" is a ranking of spatial types, in this case, in order from the highest to the lowest *i* values (Bafna, 2001). Table 4.12 shows the average connectivity analysis values of the three TCPG cases. Thus, in this case the spatial types are ranked according to the average *i* value for each type, from the highest to the lowest, as follows: pathway > small garden room > pavilion > yard > large garden room > covered corridor. However, the value differences between three of the spatial types – the pavilion (*i* = 0.92), yard (*i* = 0.89) and large garden room (*i* = 0.84 – are relatively minor. Therefore, for the purposes of this stage they have been ranked jointly to produce a simplified inequality genotype: pathway > small garden room > pavilion/yard/large garden room > covered corridor. Furthermore, in order to determine a more precise set of limits for the following stage of the research, the standard deviation (*SD*) of the *i* value was calculated to set its testing range (Table 4.13). According to the testing range, the Yuyuan garden has three spatial types within the range (pavilion, yard/square, pathway), and the Liuyuan garden has four (large room, small room, yard/square, covered corridor), while all of the space types in the Zhuozhengyuan garden are within the *i*-range tested.

4.6.2. Stage 2: Parametric generation

The basic parameters used to generate the new TCPGs for the present chapter include the number, type and mean connectivity relations (expressed using *TD, MD* and *i*) in the three historic cases. Several researchers have previously proposed using such syntactical measures as

Table 4.12. Average values of the convex space analysis of the three garden cases.

Spatial type	Number of spaces	TD	MD	i
Large room	8.67	39.00	5.19	0.84
Small room	15.00	74.67	5.44	1.56
Pavilion	10.33	53.33	5.90	0.92
Yard/Square	9.33	45.00	5.18	0.89
Covered corridor	4.67	22.00	5.72	0.35
Pathway	52.67	216.00	4.44	8.44

Table 4.13. Testing range of *i* values.

Spatial type	i	SD of i	Testing range (i-range)
Large room	0.84	0.35	0.49~1.19
Small room	1.56	0.62	0.96~2.18
Pavilion	0.92	0.51	0.40~1.43
Yard/Square	0.89	0.05	0.84~0.94
Covered corridor	0.35	0.29	0.06~0.64
Pathway	8.44	4.63	3.81~13.07

parameters. For example, Jeong and Ban (2011) developed computational algorithms to evaluate design solutions using space syntax methods. Nourian et al. (2013) have presented a computational toolkit developed for configurative architectural design using space syntax theory. Their toolkit, named "SYNTACTIC", is a plugin for Grasshopper and is useful for generating spatial diagrams and providing real-time evaluation.

Using the mathematical characteristics of the historic TCPGs as a starting point, a parametric system is authored to generate a series of spatial types and connectivity graphs that conform to these characteristics. Specifically, for each of the three historic TCPGs three new mathematically compliant plans were generated (nine in total demonstrated here). The efficacy of the generative system was determined in each of the cases by checking *TD*, *MD* and *i* against the inequality genotype and *SD* measures identified in the case studies. The average size and number of each spatial type was used to generate nodes, then connectivity points were optimised to achieve the graph values, and finally the nodes in the nine new designs transformed into abstract, but correctly scaled, spatial types. The following subsections describe how the connectivity rules were produced and the means by which the nodes were resized to replicate the spatial types.

Steps in the generative and testing process

In order to generate a parametric system that reflects the mathematical characteristics of the TCPGs, the following steps were taken.

First, some mathematical characteristics of different spatial types derived from the three TCPG cases were set as rules in Grasshopper. For instance, the pavilion typically has only one connection; a proportion of the yards are typically connected to large garden rooms, and so on. Second, the average number of each spatial type in the TCPG – such as the number of large rooms or small rooms in a garden – was then set as input parameters. Third, for this case, a pre-determined entrance and generic site boundary were selected, including the distance range of the nodes to the primary path. Fourth, new parametrically generated plans of TCPGs were produced, each suggesting a possible connectivity schema reflecting the characteristics of the historic TCPGs. Finally, the *TD*, *MD*, and *i* values of the generated parametric diagram were tested against the inequality genotypes produced for the three TCPG cases.

When performing the *i*-range testing of an exemplar parametric plan, if the *i* value was within the range (see Table 4.13), the outcome was classified as "true". The more "true" values the system generates, the more closely the characteristics of the new TCPG design reflect those of the historic cases. As the historic cases have at least three spatial types that are within the *i*-test range, in order to be confirmed as a "compliant" design, a filtering criteria was set which confirmed that the new parametric diagram must have at least three "true" values.

Figure 4.39 and Table 4.14 show an example of a parametrically generated and tested connectivity diagram for a new TCPG produced in accordance with these four steps. In this figure, different nodal symbols represent the six spatial types in the garden, along with a seventh for the entrance. The testing of the parametric diagram in the example confirms that it complies with the inequality genotype test and, in terms of the *i*-range test, four of the six properties are "true", meaning that the diagram exceeds the minimum characteristic of the TCPG.

Table 4.14. Compliance testing for the example parametric diagram.

Spatial type	Number of spaces	TD	MD	i	Inequality genotype rank	i-range test
Large room	20	24	3.43	1.24	3	False
Small room	8	64	3.37	3.80	2	False
Pavilion	8	24	3.43	1.24	3	True
Yard/Square	7	23	3.83	0.88	5	True
Covered corridor	4	12	4.00	0.33	6	True
Pathway	55	177	3.28	11.63	1	True

Parametric diagram

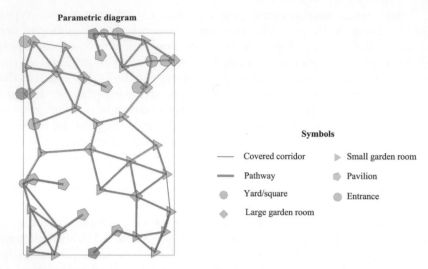

Symbols

——	Covered corridor	▷	Small garden room
▬	Pathway	⬮	Pavilion
●	Yard/square	●	Entrance
◆	Large garden room		

Figure 4.39. An example parametric diagram generated using TCPG connectivity measures.

Forming and sizing the spatial types

In order to complete the garden design, a method was required to convert the spatial-type nodes into correctly scaled, geometric forms, and the paths connecting them into practical, topographically correct routes. The first of these was achieved by comparing the size of each of the six spatial types found in the three historic cases. In this way, a range could be determined for each spatial type. Table 4.15 records the size range and common form or shape of four of the garden elements in the historic cases. The covered corridor and pathway shapes and locations in the garden plans present a different challenge. Their typical widths were determined using those in the historic cases, and their directions were prioritised to run parallel to the dominant site boundary as part of their connecting role. This is because, in the historic cases, there are two types of path forms: one is parallel with the site boundary and the other is more organic. In this study, a rule was produced to replicate the parallel pathway system only, the organic pathways being beyond the scope of the present research. With these two sets of rules implemented, the new connectivity graph was converted into a garden plan layout, with scaled and appropriately shaped elements and connections.

Parametrically generated examples

To complete this stage, for each of the three historic sites, three new compliant plans were parametrically generated (Tables 4.16–4.18). Compliant in this context means corresponding to the inequality genotype

Table 4.15. Form and size for each functional type (all units are in square metres).

	Large garden room	*Small garden room*	*Pavilion*	*Square*
Plan shape	Rectangle	Rectangle	Octagonal	Rectangle
Mean	148.47	50.17	19.55	99.42
SD	71.67	31.80	17.50	52.37
Size range	75.80~220.14	18.37~81.97	2.05~37.05	47.05~151.79

and having at least three "true" determinants in the range measures. Thus, these new plans closely model the connectivity characteristics of the historic gardens.

4.6.3. Stage 3: Fractal analysis

A fractal dimension is a non-integer value that represents the characteristic or typical distribution of detail or form in an image or object (Mandelbrot, 1982; Mandelbrot, 1977). A fractal dimension could, therefore, be thought of as a statistical measure of visual complexity or textural roughness. For a two-dimensional image, the fractal dimension (D) is a value between 1.0 and 2.0, where a higher value of D, closer to 2.0, suggests a visually complex image. Fractal dimensions have repeatedly been used since the early 1990s to analyse and compare the visual character of urban space (Batty and Longley, 1994; Chalup et al., 2009; Stamps, 2002) and architecture (Bovill, 1996; Lorenz, 2003; Ostwald et al., 2008). Significantly, fractal dimensions have also been used to analyse the Yuyuan garden (Lu, 2010) and the visual complexity of urban squares and gardens in China (Liang et al., 2013).

The standard mathematical method for calculating the fractal dimension of a two-dimensional image is presented in past research (Bovill, 1996; Lorenz, 2003; Mandelbrot, 1982). This approach, called the box-counting method, has also been substantially refined in recent years and the optimal method of application and settings for the process determined (Ostwald, 2013; Ostwald and Vaughan, 2013a, 2013b; Ostwald and Vaughan, 2016). These methodological settings and standards are used for the present research.

The three historic and nine new TCPG plans were each analysed using *ArchImage* software, and their fractal dimensions calculated (Table 4.19). For each of the new plans, the difference between its D value, and the D value of the original was calculated and presented as a percentage ($\%_{Diff}$), along with a positive or negative indicator of whether it is higher or lower (more or less visually complex) than the corresponding original. As shown in Table 4.19, there are two ways to compare these mean values. The "relative" variation includes the positive and negative values in its mean calculation. The "absolute" variation treats all differences as positives regardless of whether they are higher or lower than the target. The former is the more conventional method in fractal analysis, as it takes into account the fact that a trend in higher or lower results is potentially important to accommodate into the mean. However, for measuring the raw similarities between a target figure and variations, the latter absolute value may be more useful.

Table 4.16. Comparison of the newly generated Yuyuan Garden plans with the historical plan (spatial types are coded as per the key in Figure 4.26).

Original plan		Functional type	i	i test
		Large garden room	1.20	True
		Small garden room	2.19	True
		Pavilion	1.34	True
		Yard/Square	0.94	True
		Covered corridor	0.67	False
		Path way	12.04	True
New plan 1		Functional type	i	i test
		Large garden room	0.94	True
		Small garden room	2.04	True
		Pavilion	1.31	True
		Yard/Square	1.80	False
		Covered corridor	0.83	False
		Path way	10.54	True
New plan 2		Functional type	i	i test
		Large garden room	1.07	True
		Small garden room	3.83	False
		Pavilion	0.83	True
		Yard/Square	0.83	False
		Covered corridor	0.60	True
		Path way	12.1	True

(Contd.)

New plan 3		Functional type	*i*	*i* test
		Large garden room	0.75	True
		Small garden room	3.00	False
		Pavilion	1.25	True
		Yard/Square	2.33	False
		Covered corridor	0.43	True
		Path way	11.27	True

Table 4.17. Comparison of the newly generated Zhuozhengyuan Garden plans with the historical plan.

Original plan		Functional type	*i*	*i* test
		Large garden room	0.82	True
		Small garden room	1.53	True
		Pavilion	1.06	True
		Yard/Square	0.88	True
		Covered corridor	0.11	True
		Pathway	10.05	True
New plan 1		Functional type	*i*	*i* test
		Large garden room	0.90	True
		Small garden room	5.22	False
		Pavilion	0.70	True
		Yard/Square	1.13	False
		Covered corridor	0.56	True
		Pathway	14.50	False

(*Contd.*)

New plan 2		Functional type	*i*	*i* test
		Large garden room	0.95	True
		Small garden room	4.59	False
		Pavilion	1.12	True
		Yard/Square	0.94	True
		Covered corridor	0.71	False
		Pathway	13.4	False
New plan 3		Functional type	*i*	*i* test
		Large garden room	0.90	True
		Small garden room	6.98	False
		Pavilion	0.90	True
		Yard/Square	0.97	True
		Covered corridor	0.84	False
		Pathway	18.00	False

Table 4.18. Comparison of the newly generated Liuyuan Garden plans with historical plan.

Original plan		Functional type	*i*	*i* test
		Large garden room	0.50	True
		Small garden room	0.96	True
		Pavilion	0.34	False
		Yard/Square	0.84	True
		Covered corridor	0.27	True
		Pathway	3.22	False

(*Contd.*)

New plan 1		Functional type	*i*	*i* test
		Large garden room	0.75	True
		Small garden room	5.33	False
		Pavilion	0.63	True
		Yard/Square	1.00	False
		Covered corridor	0.43	True
		Pathway	16.15	False
New plan 2		Functional type	*i*	*i* test
		Large garden room	0.32	False
		Small garden room	3.02	False
		Pavilion	0.53	True
		Yard/Square	0.53	False
		Covered corridor	0.23	True
		Pathway	9.50	True
New plan 3		Functional type	*i*	*i* test
		Large garden room	0.71	True
		Small garden room	3.22	False
		Pavilion	0.71	True
		Yard/Square	0.71	False
		Covered corridor	0.27	True
		Pathway	10.53	True

For the relative mean, the result for the Yuyuan gardens site was 2.6% less visually complex than the original, and for the Zhuozhengyuan and Liuyuan sites, 7% less and 8.3% less respectively. Interpreting these numerical results can be undertaken in two ways. First, it is possible to

Table 4.19. Fractal dimensions (*D*) for the three original TCPGs and three new versions for each site and their relative difference from the original.

Site	Original		New version 1		New version 2		New version 3		Summary	
	D		*D*	*% Diff*	*D*	*% Diff*	*D*	*% Diff*	*Mean % Diff Relative*	*Mean % Diff Absolute*
Yuyuan	1.58		1.46	-12	1.46	-12	1.74	+16	-2.6	13.3
Zhuozhengyuan	1.55		1.48	-7	1.46	-9	1.50	-5	-7	7
Liuyuan	1.52		1.42	-10	1.43	-9	1.46	-6	-8.3	8.3

Table 4.20. Grouping of newly generated plans by fractal dimension differences, compared with compliance levels.

	Threshold	*Description*	*Generated plans*	*% Diff*	*i-test true/false (original)*
Group 1	2% - 5%	Similar	New version 3 of Zhuozhengyuan	-5	3/3 (6/6)
Group 2	6% - 10%	Comparable	New version 2 of Zhuozhengyuan	-9	3/3 (6/6)
			New version 1 of Zhuozhengyuan	-7	3/3 (6/6)
			New version 1 of Liuyuan	-10	3/3 (4/2)
			New version 2 of Liuyuan	-9	3/3 (4/2)
			New version 3 of Liuyuan	-6	3/3 (4/2)
Group 3	>10%	Dissimilar	New version 1 of Yuyuan	-12	4/2 (5/1)
			New version 2 of Yuyuan	-12	4/2 (5/1)
			New version 3 of Yuyuan	+16	4/2 (5/1)

describe a difference of less than 5% as suggesting a "similar" level of appearance, while between 5% and 10% implies a "comparable" level and above 10%, dissimilar. In this way, the absolute result for the Yuyuan gardens is similar, and the other two are comparable. Second, all three means are below the target. That is, they are consistently less complex than the original.

For the absolute mean difference, the Yuyuan gardens site result was 13.3%, the Liuyuan 8.3%, and the Zhuozhengyuan 7%. In absolute terms, all three range from being comparable to dissimilar. Furthermore, the worst absolute result, for the Yuyuan gardens, was the best of the relative results. This is because the absolute results do not provide any indication of higher or lower dimensionality. Thus, the relative result is the most useful of the two for interpreting the difference, while the absolute is a better indicator of which set of parametrically produced plans is closest to its original. For the final comparison, the new versions are classified in accordance with percentage difference and i-test results (Table 4.20).

This case demonstrates a multi-stage process for analysing an existing complex design type, then parametrically generating new instances of this type, and finally testing if these comply with both the functional and aesthetic properties of the original. This particular computational design environment has none of the standard or classical techniques typically employed when designers work with heritage structures to attempt to replicate their features. These new computational methods require a different way of thinking about design, in terms of the sequential application of formal methods of analysis, generation and testing.

4.7. Case study 5.2: Transparency and mystery in traditional Chinese private gardens

Many characteristics of architectural spaces and forms have been the subject of past research, using robust algorithmic models and data visualisation techniques. Such computational models and techniques have been optimised and developed over time, with a growing body of research successfully showing how data can be derived from architectural plans to, for example, determine pedestrian movement paths, optimise surveillance data, examine social controls and predict property values or crime rates (Bhatia et al., 2013; Ellard, 2009; Hillier, 1995; Lynch, 1960). However, the characteristics of spatial transparency and mystery in architecture and design are, by their very nature, more intangible, posing a challenge for algorithmic models. Nevertheless, researchers and designers have highlighted spatial transparency and mystery as being amongst the most important characteristics for human perceptions of safety, security and emotional wellbeing (Appleton, 1975; Hildebrand,

1999). Transparency has been associated with both the ability to visually see through space, and move through space (Rowe and Slutzky, 1963). Mystery has been associated with the limited availability of information about a space (for example via reduced capacity to see or move) or a lack of intelligibility of information (for example due to disorder or complexity in an environment) (Hillier, 1996; Ostwald and Dawes, 2013). An example of an environment which has been celebrated for these two, seemingly intangible properties is the TCPG, several of which were introduced in case study 5.1. Although some computational and mathematical models do provide measures which broadly equate to spatial transparency and mystery, they have only rarely been used to analyse these properties in design. As such, some aspects of these computational models have not yet reached the same level of maturity as those used for predicting movement, crime or property values. Nevertheless, given the cognitive importance of spatial transparency and mystery, there is a need to expand and test these existing models.

4.7.1. Pedestrian accessibility convex map analysis

Analysis of the connectivity relationships between spaces can be used to explore their physical accessibility, being the capacity of a space to be passed through. One approach to analysing connectivity is space syntax theory's convex space method. A convex space is a visually constrained or enclosed part of an environment, where all points are mutually visible to one another (Hillier and Hanson, 1984). An environment, be it a building plan or a landscaped area, can be divided into an optimal set of convex spaces (being the set of the fewest in number and largest in size, of spaces). Then these convex spaces are converted into nodes of a graph and their connections are translated into the edges of a graph. The connections are indicators of both permeability and adjacency, which refer to the availability of access between spaces (Peponis and Wineman, 2002). Together the connections (also known as edges) and spaces (also known as nodes) make up a convex map or graph of an environment (Klarqvist, 1992). Nodes in such maps do not have additional properties – like geographic location or area – they just have topological properties, although there are methods for coding additional information into the map (Eloy, 2012).

When viewed as a graph, connections between nodes represent access between two spaces and the graphs are typically illustrated using a carrier node that denotes a base space, commonly a major entryway to the location. The graph can then be drawn in a way which is analogous to a tree, where the tree's root is the carrier and the other nodes are arranged in levels based on their distance from the root. The properties of the graph can be analysed using graph theory, deriving values for integration (*i*), intelligibility (*I*), mean depth (*MD*), control value (*CV*), and total depth (*TD*). In a simple sense, such measures identify the most and least integrated spaces in a building, which are the locations where people are, all other things being equal, most or least likely to meet others. These values can also be used for comparing the properties of different spaces within a building, garden or city. For example, an important measure, in the context of the present study, is intelligibility (*I*), which is indicative of the clarity of an environment from the point of view of a user inside it (Klarqvist, 1992; Peponis and Wineman, 2002). Intelligibility provides a measure for the ease of navigation of an environment. It is mathematically defined as the Pearson correlation between integration and connectivity values of all vertices. A vertex's connectivity value is the total number of that vertex's directly connected or adjacent vertices (Hillier et al., 1987). An environment with a very low intelligibility value could be confusing, labyrinthine or mysterious. A further measure, the control value (*CV*), can be used to identify spaces in a plan which play a significant role in enabling access to new destinations. A low control value, like a low integration value, may be an indicator of an isolated, difficult to access or mysterious space.

4.7.2. Visual accessibility based isovist analysis

One important way of intuitively understanding the mathematical properties of a plan or garden is to produce a visual accessibility map of it. Such a map represents the volume of space which is visible from individual locations in the plan. Space syntax uses visual graph analysis (VGA) to generate such a map, which can be used to explore the visual properties and accessibility of the environment. One important property of each location in a plan, is the extent space which can be seen from the location, a geometric property known as an isovist (Benedikt, 1979). Since it is typically neither practical nor useful to assess all points in a plan, spaces are typically overlaid with a grid, with each cell of the grid being the approximate size of one human. Then the isovist properties from the centre of each cell are computed. A visibility graph can then be developed using grid cells as the graph's nodes, and visibility between cells as the graph's edges (Turner, 2001). As each node is an actual point in space, such a graph, unlike the convex space graph, has a low abstraction. Furthermore, since isovist graphs retain geographic properties such as location and actual size, they are considered amongst the least abstract of space syntax methodologies. Aside from the aforementioned graph-based approach, individual isovists also possess pure geometrical properties, such as perimeter, area and concavity; calculated by dividing the square of the isovist perimeter by its area's roundness (Franz and Wiener, 2008). Some isovist measurements such as Drift Magnitude, Jaggedness, Drift

Direction, and Occlusivity, are useful to distinguish between hidden or visible locations within a plan (Conroy-Dalton and Bafna, 2003). Since Jaggedness is defined by the ratio of Perimeter squared to Area, a larger Jaggedness value is indicative of a more visually complex isovist. Such a space is more complex in terms of both visual accessibility (to be seen clearly) and pedestrian navigation (to access easily). Since Occlusivity relates to the ill-defined aspects of a space's visual experience (because it is calculated as the total length of all edges which are not solid surfaces), a high Occlusivity value is indicative of greater mystery, which may also contribute to difficulties with navigation and defining spatial identity. The difference between the observer's location, and the centre of gravity of an isovist they are in, is referred to as Drift. A large difference (i.e. a large Drift value) suggests a significant "draw" or sense of directionality; a feeling like being pulled through a space. Conversely, if the Drift is very low then a space may feel unusually static. If a high correlation exists between isovist area and isovist drift, then phenomenal transparency may be highly significant in the plan. If the correlation is low, then literal transparency may be more significant. Meanwhile, an inverse correlation can be indicative of no pattern being present at all.

4.7.3. Hypothesis framing

The Yuyuan Garden in Shanghai provides the case for this study, because it is renowned for the sense of mystery and complexity it evokes. These two properties can be measured from the dual perspectives of visual accessibility and physical accessibility to assess mystery and transparency in the garden. Mysterious spaces in the Yuyuan Garden are ones where users can discover more if they proceed further into the area (Kaplan, 1988), while complex spaces are those with an excess of visual information presented (Scott, 1993). This is why mystery and complexity are commonly linked concepts within spatial studies. Based on this rationale, Table 4.21 summarises three hypotheses (and each hypothesis' mathematical indicators) which can be tested for the Yuyuan Garden.

4.7.4. Exploration of Yuyuan Garden's transparency and mystery

TCPG design principles date back to traditional Chinese culture and Daoist literature (Han, 2012; Zou, 2013), that promotes a systematic view of nature as a holistic cosmos that everything is part of. One of the key Daoist figures Zhuangzi taught that achieving a state of forgetting – both self and one's surroundings – triggered a transcendental spiritual freedom (Zhuangzi, 365-286 BC). The designers of TCPGs primarily belonged to the class of artists and educated people who decided to create small peaceful oases within larger urban plans, as a celebration of nature (Wu, 1963).

Table 4.21. Hypotheses and spatial measurements.

Hypothesis	Explanation	Mathematical indicators	Result
Mystery in the Yuyuan Garden is a by-product of structural permeability, not of visual accessibility.	Although the network of paths through the garden is difficult to comprehend and to navigate, the spaces which are seen from the paths are relatively more intelligible.	1) *Intelligibility:* the Pearson correlation between integration, and the connectivity values of the whole plan graph vertices. 2) *Control:* the extent to which a space governs access to its surrounding neighbours.	If the hypothesis is true, then: 1) *Vision-related intelligibility measures will be higher than path-related intelligibility measures* 2)*Vision-related control measures will be higher than path-related control measures.*
The complexity of the Yuyuan Garden's spaces visually heightens its sense of mystery.	Visual complexity and visual mystery are both expected to be high in the garden's spaces.	1) *Occusivity:* total length of all of the occluded edges within a view-shed. 2) *Jaggedness:* ratio of perimeter squared to area within a view-shed.	If the hypothesis is true, then: A positive correlation exists between spatial integration (i) and both jaggedness (J) and occlusivity (O).
The Yuyuan Garden's phenomenal transparency effect is a by-product of the directionality and visual pull experienced in major spaces.	In large spaces, the sense of spatial transparency is due to spatial irregularity, resulting in the eye being drawn to gaze deeper into the environment.	1) Isovist area: the isovist polygon's area. 2) *Drift Magnitude:* distance from the observer's location point, to the centre of mass of the isovist polygon.	If the hypothesis is true then: *Drift magnitude increases in parallel to isovist area.*

Such circumstances led to the TCPGs typically being relatively small in area, although their skilful creation and wandering pathways, left visitors feeling that the spaces were much larger. TCPG's wandering pathways often led people to unexpected destinations, analogous to travelling through a lifelong journey (Zou, 2013). In such a way they were able to provide a simulation of travel to remote places, including the sense of being lost first, then discovering new areas, and in this process creating an experiential journey. There are also parallels with the traditional Feng Shui interpretation, that a zigzagging movement can result from the cosmological spatial-numerical diagram and natural topography.

The Yuyuan Garden is one of the most famous of all TCPGs. It was built in the 16th century, is located at the heart of southern China's city of Shanghai, and has a garden area of approximately 20,000 square metres. Although portions of the garden were destroyed in WWII, most of those sections have since been rebuilt. It is a modest scale garden, with a spatial configuration that encapsulates the typical spatial characteristics of the classical TCPG. Yuyuan Garden is famous for its subtle planning and artificial landscape with water at the centre. Prior studies have highlighted this garden in the context of debates pertaining to mystery and transparency, for example suggesting that the garden enhances its spatial experience through the lengthening of walkways and addition of visual blocks (such as exterior walls and plants) to deliberately enhance its feeling of mystery (Gu, 2013).

Hypothesis 1: *The Yuyuan Garden's mystery is a by-product of structural permeability, not of visual accessibility.*

A TCPG's planning can be understood either by way of the visually accessible aspects of its plan, or through consideration of the pedestrian accessibility of its plan. This study produced an integration map of both of these aspects of the plan, with the goal of exploring the sense of mystery in the Yuyuan Garden. VGA map data and convex map data (based on connectivity of spaces available for entering) were both utilised by this study as input data for *UCL Depthmap* software. For the former, the visible spatial boundaries served as input data, while for the latter the initial input data were both the links between spaces and the spaces themselves.

Heat maps, with a colour gradient from red to blue (dark blue signifying lowest values, and red signifying highest values) were generated in *UCL Depthmap*. Figure 4.40 (a) is an integration heat map for the pedestrian accessible portions of the garden, with red colour showing the pathways which are most statistically likely to be used by people based solely on accessibility (i.e. highest integration values). Conversely the dark blue paths in Figure 4.40 are statistically least likely to be used by people based on accessibility. The figure illustrates that from the perspective of pedestrian accessibility, the southern area and northeast corner are least integrated, and the centre area around the water is the most integrated part. Figure 4.40 (b) is an integration heat map for visually accessible portions of the garden. The figure shows that the least transparent areas of the garden plan are in the south corner area that is blocked by solid walls and the northwest area behind the hill; while the most transparent area of the garden plan is its central area.

Figure 4.40. (a) Integration heat map of Yuyuan Garden's pedestrian accessibility and (b) integration heat map of Yuyuan Garden's visual accessibility.

Due to their close connection with spatial navigation, two further variables – control and intelligibility, were measured, to explore openness and mystery in Yuyuan Garden. Control is defined as the degree by which a space is limited in its spatial access to neighbours, which is associated with the relative impact of the space on navigation when a person is traversing through that space. Intelligibility is indicative of how challenging people find navigation through a space. The values of control and intelligibility for both visual accessibility and pedestrian accessibility of Yuyuan Garden are in Table 4.22. The data shows that control is higher for visual accessibility than pedestrian accessibility, which is suggestive of enhanced mystery along the walking paths. Similarly, intelligibility is also higher for visual accessibility than pedestrian accessibility, meaning it is easier to comprehend visuality, which also implies that the sense of mystery is associated with movement. Such data and visualisations provide an import method for understanding the properties of real world cases. In the Yuyuan Garden, people find it more difficulty to travel to spaces than to see them.

Table 4.22. Yuyuan Garden's intelligibility and control results, for convex and VGA maps.

	Intelligibility	*Control*
Pedestrian accessibility	0.570	0.929
Visual accessibility	0.763	1.000

Hypothesis 2: *The complexity of the Yuyuan Garden's spaces boost its visual sense of mystery.*

The three measurements of integration, isovist occlusivity and isovist jaggedness were used to test this hypothesis. Figure 4.41 illustrates three zones of the Yuyuan Garden, which were identified in the previous stage, as being worthy of further analysis. Zone 1 is at the heart of the red-circled high integration zone, while zones 2 and 3 are in blue-circled low integration zones. Figure 4.42 shows each of the three zones' isovist polygon areas, and demonstrates that the high integration zone has a

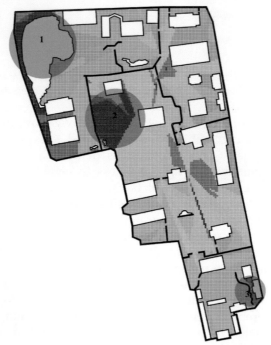

Figure 4.41. VGA map illustrating zones 1, 2 and 3 of Yuyuan Garden.

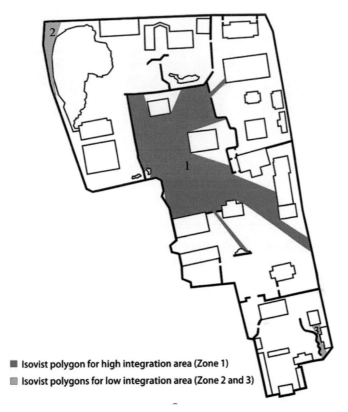

■ Isovist polygon for high integration area (Zone 1)

▨ Isovist polygons for low integration area (Zone 2 and 3)

Figure 4.42. Isovist polygons illustrating zones 1, 2 and 3 of Yuyuan Garden.

larger isovist area compared with the two low integration zones. That is suggestive of the low integration space being likely to be visually concealed by hills, buildings or walls; meanwhile the high integration space has an open, visually permeable character within a larger area.

Figure 4.43 illustrates the isovist area, isovist occlusivity and isovist perimeter heat maps for the Yuyuan Garden. Table 4.23 shows the jaggedness values (level of visual complexity from a space) and occlusivity values (the ill-defined spatial properties, which are potential sources of confusion) for the three designated zones analysed in the previous step. A correlation between high integration and high occlusivity and jaggedness is demonstrated in Table 4.23, implying that this garden's significant spaces (i.e. high *i* value) are also its most mysterious (high *O* value) and spatially complex (high *J* value). In other words, Yuyuan Garden's most complex and mysterious spaces are also its most important in terms of navigation and social co-presence.

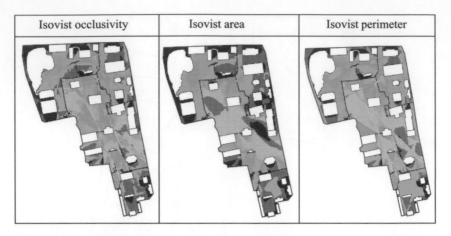

Figure 4.43. Heat maps of Yuyuan Garden's isovist area, isovist occlusivity, and isovist perimeter.

Table 4.23. VGA zone results of Yuyuan Garden's isovist occlusivity and jaggedness.

Isovist	Occlusivity (O)	Jaggedness (J)
Zone 1	273.657	84.726
Zone 2	23.063	61.356
Zone 3	24.278	40.074

Hypothesis 3: *The Yuyuan Garden's phenomenal transparency effect is a by-product of the directionality and visual pull experienced in major spaces.*

Transparency, as a spatial property, is calculated using two measures. First, isovist drift magnitude measures the extent to which people are drawn through space, a property also known as *phenomenal transparency*. Second, isovist area is a measure of the extent of the visible portion of the space, which correlates to *literal transparency*. Studies indicate that although very large spaces offer extensive visible areas they could nevertheless be uninteresting, whereas a smaller space could conceivably draw a viewer's gaze more deeply into the environment, meaning it is possible for a smaller space to have higher overall transparency (Rowe and Slutzky, 1963; Ellard, 2009). Examining the red and yellow areas in Figure 4.44's VGA visualisation of isovist drift magnitude and isovist area, shows that isovist area is more differentiated throughout the garden than isovist drift magnitude. This would mean that phenomenal transparency is more evenly dispersed throughout the Yuyuan Garden than literal transparency is. Examining Table 4.24's results for the three aforementioned zones, shows that as isovist area and drift increase, so does integration, however its correlation is non-linear. Thus, for isovist drift magnitude, zone 1's result is 5.01 times greater than for zone 2, but zone 1 has 17.76 times larger isovist area than zone 2. Similarly, zone 2's isovist drift magnitude is 6.08 times greater than for zone 3, but zone 2 has 3.59 times larger isovist area than zone 3. Overall the heat maps reveal that both literal transparency and phenomenal transparency contribute to shaping the Yuyuan Garden's spatial experience, however the latter is more diverse than the former. In other words, literal transparency only dominates spatial experience in some specific areas of the garden, while phenomenal transparency (a person's sense that further passage is possible) consistently shapes this TCPG's transparent spatial properties throughout most of the garden.

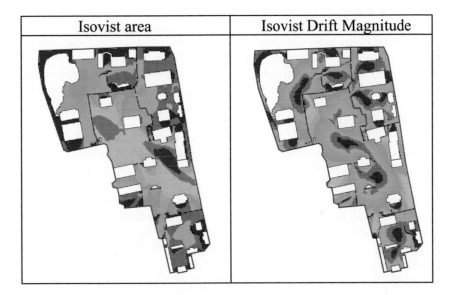

Figure 4.44. Heat maps of Yuyuan Garden's isovist drift magnitude and isovist area.

Table 4.24. VGA zone results of Yuyuan Garden's isovist drift magnitude and isovist area.

Isovist	Area (A)	Drift magnitude (D_M)
Zone 1	2739.676	23.416
Zone 2	154.225	4.666
Zone 3	42.956	0.684

In addition to the specific research findings about the Yuyuan Garden, and the application of alternative readings of space syntax measures and techniques, this case study shows that in order to support the analysis and comparison of complex data sets, computational environments will need to possess an increased capacity for data visualisation as well as design analysis and generation.

4.8. Case study 6: Creativity in a parametric design environment

The final case study is concerned with Parametric Design Environments (PDEs) and their potential links to creativity. Past research argues that parametric design is a new mode of thinking that allows designers to explore the entirety of the design space within the identified constraints (Hesselgren et al., 2007). In parametric design, the decomposition of design into rules, and the flexibility of allowing more holistic changes to the design, is also allegedly beneficial for creative results (Wortmann and Tunçer, 2017). For example, Jabi et al. (2017) believe that parametric design might be appropriate for addressing problems with a high level of complexity, thereby encouraging designers to explore the unexpected in the design solution space.

In order to understand creativity in a parametric design environment, Iordanova et al. (2009) observed a small group of architectural students using parametric software, following the design process from generation of ideas through to exploration and evolution of ideas using three paradigms: direct description, formulation and generation. They assessed the outcomes in terms of abundance, flexibility, evolution and originality of ideas, as well as efficiency and coherency. Their study linked abundance to creativity in parametric design, especially with the generative method because new ideas are emerging simultaneously as variations. This correlates to Abdelsalam's (2009) claim that PDEs support "designers to direct their creativity to a wide range of exploration using 'what-if' scenarios" (Abdelsalam, 2009, p. 299). Furthermore, during Iordanova et al.'s experiment, students saw the production of unforeseen results as a means of promoting design creativity. In contrast, Schnabel (2007) studied creativity in PDEs using observations of 30 students undertaking a more complex design task for a longer duration. Unlike the results of the other studies, Schnabel argues that parametric design provides opportunities for in-depth comprehension of design objectives, opening up a "novel set of opportunities" (p. 247). He is, however, less emphatic about the claimed relationship between creativity and parametric design. Lee et al. (2014) also explored creativity in PDEs using protocol analysis and the Consensual Assessment Technique (CAT) (Amabile, 1982). In Lee et al.'s study, three designers' parametric design processes were analysed in detail using protocol analysis and the results evaluated by experts. The study suggests that "analysis" and "synthesis" are positively related to creativity in parametric design. However, the research did not benchmark the findings against a baseline indicator (a traditional design or geometric environment).

In contrast, other researchers (e.g. Salim and Burry, 2010) believe that, in some circumstances, parametric design can actually hinder creativity. This is because, with the number of potential parametric variations increasing, flexible changes can be unmanageable. They argue that parametric design tools have certainly assisted creative thinking through rule algorithms. However, it can also decrease design possibilities because of fixed variations. Furthermore, "the more complex the software is, the lower the usability" (Salim and Burry, 2010, p. 490). In order to innovate, parametric designers need to explore their ideas using complementary creative methods (not necessarily parametric).

The last case in this chapter uses protocol analysis, correlated to the results of expert evaluation of design outcomes, to compare creativity in two different design environments. For this final study, eight designers each produced a concept design in a Parametric Design Environment (PDE) and also in a Geometrical Modelling Environment (GME), leading to 16 designs which are assessed.

4.8.1. Research design

To assess levels of creativity, the outcomes produced by eight designers, each producing a design in two environments, were displayed in an online survey, and 19 experts were invited to assess the 16 designs according to the given criteria (with scores ranging from 0–7 in each sub-category). Experts were recruited via email invitation from university and senior, experienced postgraduates. The order in which the designs were presented to the experts was randomised. Each was presented as an image accompanied by a paragraph of description. Figure 4.45 shows an example web page from the online evaluation, which presents one of the designs for expert assessment. There are three scores and a combined score: *innovation, usefulness, surprisingness* and *overall creativity*. Innovation occurs when a design encapsulates something new, original or never seen before (Sarkar and Chakrabarti, 2011; Sternberg and Lubart, 1999). Usefulness is a measure of the capacity of a design to appropriately solve a problem (Nguyen and Shanks, 2009; Nickerson, 1994; Sternberg and Lubart, 1999) Surprisingness is a measure of unusualness or unexpectedness (Jackson and Messick, 1965). In combination, overall creativity is a measure of the way a design combines and encapsulates innovation, usefulness and surprisingness. Definitions of these were provided to the expert assessors.

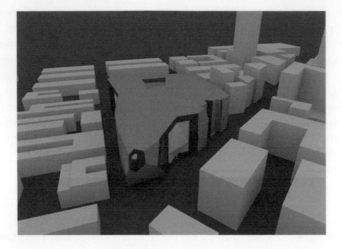

- This concept design is for a community centre on an urban site.

- The designer quickly defined a building mass using the site boundary.

- For the elevation design, they treat each façade as segments and layers of a tree. During the modelling, they sliced the façade layer by layer, in order to set the shape, and adjusted the angles.

- Their intention was to create an irregular and interesting façade.

What do you think of this design?

	1 (Minimal level)	2	3	4	5	6	7(Maximal level)
Overall creativity	○	○	○	○	○	○	○
Innovation	○	○	○	○	○	○	○
Usefulness	○	○	○	○	○	○	○
Surprising	○	○	○	○	○	○	○

Figure 4.45. An example web page of the online evaluation.

4.8.2. Analysis of results

Table 4.25 shows selected illustrations of the design outcomes collected from the experiment. Table 4.26 shows the jury evaluation results of these outcomes. The results indicate that the average overall creativity score of designs produced in PDE (4.11) is 4.62% higher than those produced in GME (3.92). For *innovation*, the average score in PDE (3.84) is 5.47% higher than its counterpart's GME score (3.63); for *surprisingness*, PDE (3.93) is 9.16% higher than GME (3.57), and finally for *usefulness*, interestingly, the average score in GME is 1.27% higher than the one in PDE. From these results we might infer that PDEs can potentially support creativity

Table 4.25. Design outcomes collected from the design experiment.

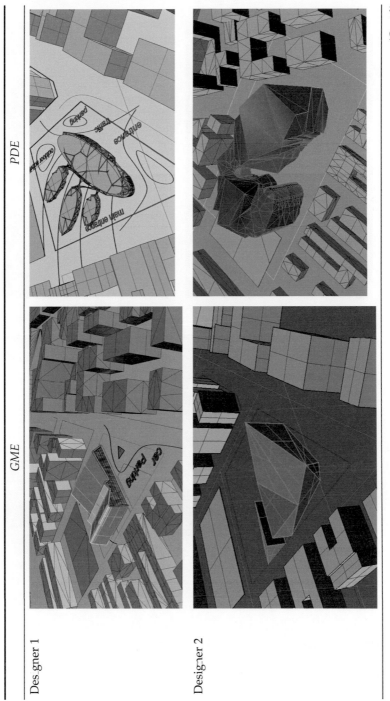

(Contd.)

Table 4.25. (*Contd.*)

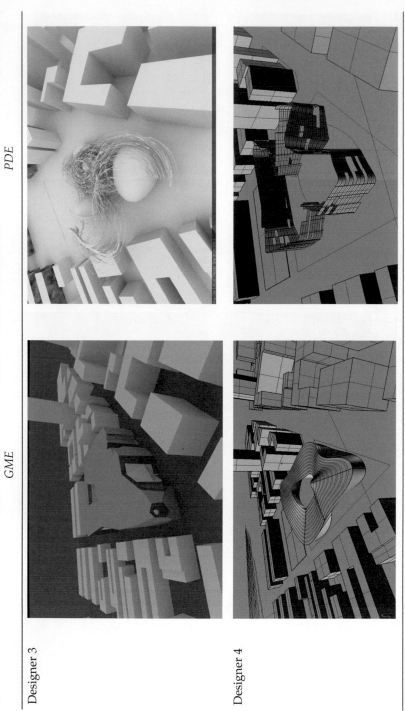

GME
PDE
Designer 3
Designer 4

(*Contd.*)

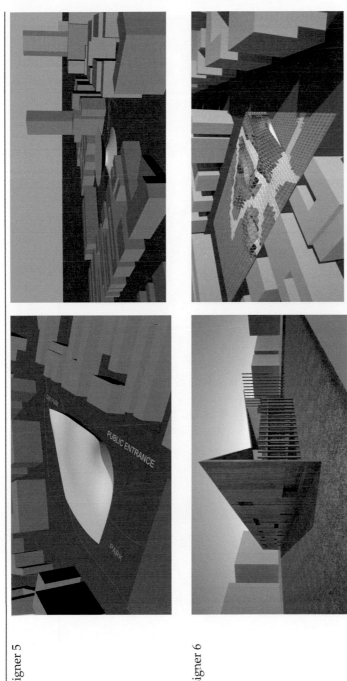

(Contd.)

Table 4.25. (*Contd.*)

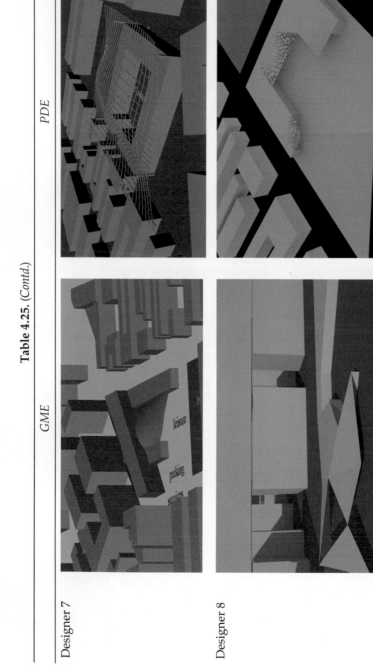

in terms of overall creativity, innovation and surprisingness especially. This may be due to the form-making capabilities of parametric design in comparison with traditional geometric modelling. However, PDEs may be less effective for supporting useful design outcomes than GMEs.

Table 4.26. Jury evaluation results of the collected design outcomes.

	Average of jury evaluation results	Overall creativity	Innovation	Usefulness	Surprisingness
GME	Designer 1	3.58	3.16	3.53	3.11
	Designer 2	4.05	3.58	3.95	3.47
	Designer 3	3.89	3.68	3.74	3.78
	Designer 4	4.32	4.21	4.22	4.11
	Designer 5	3.95	3.84	3.37	3.79
	Designer 6	3.95	3.58	4.74	3.53
	Designer 7	3.05	2.95	3.53	2.79
	Designer 8	4.58	4.05	4.32	4.00
PDE	Designer 1	4.58	4.11	4.26	3.68
	Designer 2	3.95	3.68	3.79	3.68
	Designer 3	4.42	4.05	3.16	4.79
	Designer 4	4.05	3.74	3.11	3.79
	Designer 5	3.11	3.21	2.53	3.47
	Designer 6	4.89	4.53	5.00	4.63
	Designer 7	3.89	3.79	4.84	3.63
	Designer 8	3.95	3.63	4.37	3.74
GME	Average	3.92	3.63	3.93	3.57
	SD	0.461	0.422	0.472	0.447
PDE	Average	4.11	3.84	3.88	3.93
	SD	0.538	0.391	0.887	0.495
Differences		4.62%	5.47%	1.27%	9.16%

To review the results, they are grouped into the three highest and three lowest evaluation scores in each design environment in terms of *overall creativity, innovation, usefulness* and *surprisingness* (Table 4.27). For instance, Designers 4, 8 and 2 are grouped into the highest overall creativity group in GME, while Designers 1, 3 and 7 are in the lowest overall creativity group in GME. Then using the protocol data from the experiments as presented in case study 1, the corresponding coding results of design process (design issue distribution) for high and low evaluation score groups in both design environments are compared. Table 4.28 shows the pairings in term of *overall creativity*. Since there are not enough segments in Requirement (R) to produce statistically valid results, only F, Be, Bs and S are analysed. Differences between the high and low score groups in each design environment are calculated and compared. The threshold is set for a 30% difference as statistically significant. In Table 4.28 it is apparent that there is no significant difference in the two groups between the two design environments.

Table 4.27. Grouping of high and low evaluation score groups.

		GME	PDE
Overall creativity	High score group	Designer 4, 8, 2	Designer 6, 1, 3
	Low score group	Designer 7, 1, 3	Designer 5, 7, 8, 2
Innovation	High score group	Designer 4, 8, 5	Designer 6, 1, 3
	Low score group	Designer 7, 1, 2, 6	Designer 5, 2, 8
Usefulness	High score group	Designer 4, 6, 8	Designer 6, 7, 8
	Low score group	Designer 5, 7, 1	Designer 5, 4, 2
Surprising	High score group	Designer 4, 8, 5	Designer 3, 6, 4
	Low score group	Designer 7, 1, 2	Designer 5, 7, 1, 2

Table 4.28. Design issue distribution results for both high and low evaluation score groups in terms of *overall creativity*.

		F (%)	Be (%)	Bs (%)	S (%)
GME	High score	6.83	17.96	34.13	39.34
	Low score	8.86	20.52	26.95	38.90
	Difference	−22.87%	−12.48%	21.03%	1.13%
PDE	High score	4.97	15.73	29.26	48.64
	Low score	5.62	27.26	28.01	37.44
	Difference	−11.55%	−42.28%	4.29%	23.03%
PDE vs. GME	Difference between high and low group	11.32%	29.8%	16.74%	21.9%

In terms of innovation, in Table 4.29 the difference in Expected Behaviour (Be) between the PDE (−35.74%) and the GME (6.71%) is significant (42.45%). In the PDE, *Be* is higher in the low evaluation score group (24.48%) than the higher evaluation score group (15.73%). This suggests that in the PDE, designers' activities related to *Be* tend to hinder "innovation" while this is not the case of GME. This may be because designers need to allocate extra effort to setting rule algorithm related goals in the PDE, which result in lesser time and effort consumption to attain design related goals. This could lead to the outcomes being less innovative.

Table 4.29. Design issue distribution results for both high and low evaluation score groups in terms of *innovation*.

		F (%)	Be (%)	Bs (%)	S (%)
GME	High score	7.60	19.75	31.98	38.50
	Low score	9.11	18.43	28.10	41.11
	Difference	−16.59%	6.71%	12.13%	−6.36%
PDE	High score	4.97	15.73	29.26	48.64
	Low score	4.49	24.48	31.04	38.36
	Difference	9.65%	−35.74%	−5.71%	21.14%
PDE vs. GME	Difference between high and low group	26.24%	42.45%	17.84%	27.5%

In terms of usefulness, the results in Table 4.30 show that the differences in Function (F), Expected Behaviour (Be) and Structure Behaviour (Bs) between the PDE and the GME are significant, which are 96.04%, 42.14%, and 38.24% respectively. This suggests that these three issues could have a very significant impact on the design outcome in terms of *usefulness*. In the PDE, Function is significantly higher (54.35%) in the high evaluation score group (8.05%) than the low evaluation score group (3.68%).

Table 4.30. Design issue distribution results for high and low evaluation score groups in terms of *usefulness*.

		F (%)	Be (%)	Bs (%)	S (%)
GME	High score	6.35	17.84	30.29	44.33
	Low score	10.89	22.17	26.79	35.37
	Difference	−41.69%	−19.56%	11.55%	20.23%
PDE	High score	8.05	30.43	23.18	36.78
	Low score	3.68	23.56	31.62	40.05
	Difference	54.35%	22.58%	−26.69%	−8.16%
PDE vs. GME	Difference between high and low group	96.04%	42.14%	38.24%	28.39%

In terms of Surprisingness, Table 4.31 shows that the difference of Be between the PDE (−3.09%) and the GME (−37.70%) is significant (34.61%). This suggests that designers consider Be differently when designing in a PDE and a GME, which can have an impact on their design outcomes in terms of *surprisingness*. In both environments, the low evaluation score group has a higher Be, which might mean that this type of activity can potentially hinder the creativity in design outcomes in terms of *surprisingness*.

Table 4.31. Design issue distributions for both high and low evaluation score groups in terms of *surprisingness*.

		F (%)	Be (%)	Bs (%)	S (%)
GME	High score	22.79	59.25	95.94	115.49
	Low score	30.38	61.14	86.82	108.63
	Difference	−24.98%	−3.09%	9.51%	5.94%
PDE	High score	14.19	58.91	89.70	133.97
	Low score	20.77	94.56	111.88	166.77
	Difference	−31.68%	−37.70%	−19.82%	−19.67%
PDE vs. GME	Difference between high and low group	6.7%	34.61%	29.33%	25.61%

4.9. Conclusion

This chapter has explored the impact of computational design environments on design and designers via six case studies. These studies examined six distinct design environments (TDE, VDE, GME, PDE, GDG, GDE) and three variations of synchronous collaborative environments. Key conclusions and observations arising from each are summarised hereafter.

Case study 1

The primary conclusion arising from this study is that the PDE does not necessarily change a designer's higher-level thinking. The factors that do not change include the effort invested by the designer in each design issue, the tendencies of their design moves and the way they explore problems and solutions. There is a difference, however, in the way designers' cognitive effort is allocated to tasks. Specifically, in the GME designers expended most of their cognitive effort at the design knowledge level. At this level, designers consider issues such as how to adapt a building to the site, how to shape the way people use a building, and how to satisfy the requirements of clients. In the PDE, most of the time was spent supporting geometric modelling. As Burry states, in most cases when using parametric tools the variables are "those that define the measurement of entities and distance along with their relative angles" (Burry, 2003, p. 211), which indicates that the main variables in parametric design focus on geometrical elements. The other difference between the

PDEs and GMEs is related to the application of prior knowledge, which includes both the experience of architectural design and the experience of using parametric tools. In PDEs, designers tend to use the scripts they are familiar with, and adapt them to the current design context. Such scripts can make parametric design tools more efficient.

Case study 2

Results of this study confirm that 3D virtual worlds can support synchronous design collaboration between designers who are remotely located without major compromises in design communication and representation. This suggests that collaboration in Virtual Design Environments (VDEs) is potentially as effective as co-located and remote design in TDEs. The study further indicates that a multi-user virtual world that allows co-creation and co-editing of 3D models changes the behaviour of designers in two important ways. First, designers worked collectively on the same task for most of the time while co-located in the TDE and the virtual world. In contrast, they only worked together for part of the time when remotely located in different physical locations. Second, whereas sketches were sequentially produced and abandoned in co-located and remote versions of the TDEs, in the VDE the 3D model was the constant focus of development and communication. These results show that designing with 2D sketches in TDEs and 3D models in VDEs can change designers' cognitive and behavioural approaches.

Case study 3

The results of this study demonstrate that the use of eye-tracking in design research can connect biometric evidence to cognitive design data. Specifically, whereas past research suggests that people focus their visual attention on edges rather than surfaces (Weber et al., 2002), this study showed that in a GME, designers focused their attention on façades rather than edges or corners. This might suggest that CAD tools should have improved functionality for both space and shape making, rather than just shape making. Furthermore, contrary to past research (Marr, 2010), designing is a generation process rather than an object recognition process, and a person's eye movement patterns will be different for these tasks (Cyganek, 2013; Ullman, 2000). A further observation is that constantly checking the 3D model is essential during the design process (Yu et al., 2013). In addition, experienced designers rarely focused on the GME menu or interface. Finally, most of the designers' gazes focused on the middle of the screen or towards the left. Marr (2010) suggests that 2D sketching is more viewer-centred in terms of geometry, while 3D images are more object-centred, which includes volumetric primitives. These results provide the foundation for hypotheses that can be tested in more generalisable experiments.

Case study 4

The results of this study demonstrate that the integration of VDEs and Generative Design Grammars (GDG) can go beyond the conventional purpose of design communication and static simulation to support generation and automation of design. The GDG provides an integrated framework and powerful agent reasoning and simulation that can automate and optimise parts of the design process. More importantly, the GDG serves as a platform for representing design requirements as well as simulating contextual information that may affect the generation of a design. In parallel, the use of Generative Design Agents (GDA) presents a robust approach to design reasoning, generation and automation. These design agents actively seek to satisfy their design goals to meet the changing requirements by interacting with the environment (Wooldridge and Jennings, 1995). The use of GDA in conjunction with a GDG offers a robust approach to design reasoning and automation. The effectiveness of both GDG and GDA is demonstrated in this study through a virtual design scenario for a gallery. Although the design scenario is constructed with a specific kind of gallery in mind, and a specific platform of a VDE, it confirms the effectiveness of GDG and GDA. Integrated with different design and domain knowledges, they can be adapted for dynamic and autonomous design in 3D virtual environments for other purposes. The GDG framework developed in this study also provides a foundation to formally study different types of virtual environments.

Case study 5

Case study 5 encompasses two garden studies. Case study 5.1 demonstrates a multi-component Generative Design Environment (GDE) for analysing a body of work, extracting its characteristics, and generating new works from these rules. Specifically, it uses parametric design to generate new garden plans that not only conform to selected socio-spatial patterns present in 16th century TCPGs, but also have broadly similar levels of visual richness and complexity. Through this research an important new application of generative design is proposed, along with a potential new insight into the visual and geometric character of TCPGs. In this study, the GDE is essentially a series of computational operations, some analytical, some generative. This is a flexible system that is appropriate for producing new plan variants at the conceptual design stage. When multiple designs are generated, based on the parametric rule sets, the plans comply with the character of its original site or context, leading to potential applications for designers in multiple fields. The second observation of the study is that the parametric system was able to generate garden plans which are visually similar or comparable to the original TCPGs using only connectivity measures. This implies that a large part of the much-celebrated visual character of the TCPG may be derived from the structure of its network of spaces.

Case study 5.2 draws on computational analysis and visualisation techniques to analyse two properties of a complex garden environment. These properties are both associated with transparency, which is defined in terms of capacity for movement and view. Through this methodology new insights are revealed into two perceptual properties – mystery and transparency – which have been widely attributed to TCPGs, yet are insufficiently understood. The findings in this chapter indicate that Yuyuan Garden's quality of spatial mystery is less a result of its visual accessibility, and more a result of its structural permeability, which is further reinforced and amplified by this garden's spatial complexity. These results also conclude that this garden's overall transparency is a property which mostly results from the visual pull or directionality of the garden's major spaces. This study's three main contributions have been to: i) further our understanding of TCPG spatial features, ii) test and establish a space syntax methodology for examining spatial features of complex landscape designs, and iii) demonstrate that space syntax visualisation techniques are well suited for intuitive presentation of such results. To effectively support design and analysis, computational environments should play a key role in data visualisation in addition to the standard design analysis and generation roles.

Case study 6

The results of the final study indicate that a PDE can potentially assist the emergence of innovation, surprisingness and overall creativity in design. This finding broadly correlates with the results of previous research, which argues that parametric design can lead to an abundance of options, and thereby produce unforeseen or unpredictable design results (Iordanova et al., 2009; Schnabel, 2007). There were however, no significant differences observed between the two design environments in terms of perceived usefulness or functionality of outcomes. There were also no significant differences in the distribution of design issues between the high and low evaluation score groups, in terms of overall creativity in both design environments. The distributions of Expected Behaviour (Be) did however, differ significantly between the PDEs and the GMEs, in terms of the innovation, usefulness and surprisingness. Thus, *Be* activities in PDEs may influence perceptions of innovation, usefulness and surprisingness in design outcomes. Further, the distributions of Function (F) and Structure Behaviour (Bs) are also significantly different between the PDE and the GME for the evaluation of usefulness, especially for *F* Emphasising rule-algorithm related functions in the PDE may supports design creativity by producing more useful design solutions. For example, the use of building performance related rule-algorithm settings during parametric design might create more original or unexpected, but still useful, design solutions.

References

Abdelsalam, M. (2009). The use of the smart geometry through various design processes: Using the programming platform (parametric features) and generative components. *In: Proceedings of the Arab Society for Computer Aided Architectural Design (ASCAAD 2009)*, Manama, Kingdom of Bahrain. http://papers.cumincad.org/cgi-bin/works/paper/ascaad2009_mai_abdelsalam

Alexander, C. (1979). *The Timeless Way of Building*. Oxford University Press. http://books.google.com.au/books?id=H6CE9hlbO8sC

Alexander , C., Ishikawa, S. and Silverstein, M. (1977). *A Pattern Language: Towns, Buildings, Construction*. Oxford University Press.

Aish, R. (2005). From intuition to precision. *Proceedings of 23rd eCAADe Conference*, pp. 10-14, Lisbon, Portugal.

Amabile, T.M. (1982). Social psychology of creativity: A consensual assessment technique. *Journal of Personality and Social Psychology*, 43, 997–1013.

Appleton, J. (1975). *The Experience of Landscape*. John Wiley and Sons.

Arnheim, R. (1974). *Art and Visual Perception: A Psychology of the Creative Eye*. University of California Press.

Bafna, S. (2001). Geometric intuitions of genotypes. *Proceedings of the Third International Symposium on Space Syntax*.

Bafna, S. (2003). Space syntax: A brief introduction to its logic and analytical techniques. *Environment and Behavior*, 35(1), 17–29.

Batty, M., and Longley, P. (1994). *Fractal Cities: A Geometry of Form and Function*. Academic Press.

Benedikt, M.L. (1979). To take hold of space: Isovists and isovist fields. *Environment and Planning B: Planning and Design*, 6(1), 47–65.

Bhatia, S., Chalup, S.K. and Ostwald, M.J. (2013). Wayfinding: A method for the empirical evaluation of structural saliency using 3D Isovists. *Architectural Science Review*, 56(3), 220–231.

Bovill, C. (1996). *Fractal Geometry in Architecture and Design*. Birkhauser.

Burry, M. (2003). Between intuition and process: Parametric design and rapid prototyping. pp. 149–162. *In*: B. Kolarevic (Ed.), Architecture in the Digital Age—Design and Manufacturing. Spon Press.

Chalup , S.K., Henderson, N., Ostwald, M.J. and Wiklendt, L. (2009). A computational approach to fractal analysis of a cityscape's skyline. *Architectural Science Review*, 52(2), 126–134.

Chang, H.-y. (2006). The Spatial Structure Form of Traditional Chinese Garden—A Case Study on The Lin Family Garden [Master's thesis]. National Chenggong University, Taiwan.

Chen, R. (2012). Research on Traditional Chinese Garden's Space Syntax and Insights into Contemporary Regional Reconstruction [Master's thesis]. Tsinghua University.

Chi, M.T.H. (1997). Quantifying qualitative analyses of verbal data: A practical guide. *Learning Science*, 6(3), 271–315.

Chien, S.-F. and Yeh, Y.-T. (2012). On creativity and parametric design—A preliminary study of designer's behaviour when employing parametric design tools. pp. 245–253. *In*: H. Achten, J. Pavlíček and J Huhín (Eds.), *Proceedings of 30th International Conference on Education and Research in Computer Aided Architectural Design in Europe (eCAADe 2012)*. Czech Republic. http://papers.cumincad.org/cgi-bin/works/Show?ecaade2012_223.

Ching, F. (2014). *Architecture: Form, Space, and Order* (4th ed.). Wiley.

Ching, W.K. and Ng, M.K. (2006). *Markov Chains: Models, Algorithms and Applications*. Springer.

Conroy-Dalton, R. and Bafna, S. (2003). The syntactical image of the city: A reciprocal definition of spatial elements and spatial syntaxes. *Proceedings of the 4th International Space Syntax Symposium*, London.

Corne, D., Smithers, T. and Ross, P. (1993). Solving design problems by computational exploration. *In*: J. Gero and N. Tyugy (Eds.), Formal Design Methods for Computer-aided Design. Elsevier, New York.

Cross, N. (2004). Expertise in design: An overview. *Design Studies*, 25(5), 427–441. https://doi.org/10.1016/j.destud.2004.06.002

Cross, N. (2011). *Design Thinking: Understanding How Designers Think and Work* (English ed.). Berg Publishers.

Cyganek, B. (2013). *Object Detection and Recognition in Digital Images: Theory and Practice*. Wiley.

Darken, R.P. and Sibert, J.L. (1993). A toolset for navigation in virtual environments. pp. 157–165. *In: Proceedings of the 6th Annual ACM Symposium on User Interface Software and Technology*. Atlanta.

Darken, R.P. and Sibert, J.L. (1996). Wayfinding strategies and behaviours in large virtual worlds. pp. 142–149. *In: CHI '96: Proceedings of the SIGCHI Conference on Human Factors in Computing Systems (ACM SIGCHI'96)*. https://doi.org/10.1145/238386.238459

Dawes, M.J. and Ostwald, M.J. (2017). Christopher Alexander's A Pattern Language: Analysing, mapping and classifying the critical response. *City, Territory and Architecture*, 4(1), 17. https://doi.org/10.1186/s40410-017-0073-1

Dawes, M.J. and Ostwald, M.J. (2020). The mathematical structure of Alexander's A Pattern Language: An analysis of the role of invariant patterns. *Environment and Planning B: Urban Analytics and City Science*, 47(1), 7–24. https://doi.org/10.1177/2399808318761396

Dorst, K. and Cross, N. (2001). Creativity in the design process: Co-evolution of problem-solution. *Design Studies*, 22(5), 425–437. https://doi.org/10.1016/s0142-694x(01)00009-6

Duarte, J. (1999). Democratized architecture: Grammars and computers for Siza's mass housing. *In: Proceedings of the International Conference on Enhancement and Promotion of Computational Methods in Engineering and Science*. Macau.

Ellard, C. (2009). *You Are Here*. Random House.

Eloy, S. (2012). *A Transformation Grammar-based Methodology for Housing Rehabilitation: Meeting Contemporary Functional and ICT Requirements* Lisbon: University of Technology.

Fowler, M. (2003). *Patterns of Enterprise Application Architecture*. Addison-Wesley. http://books.google.com.au/books?id=FyWZt5DdvFkC

Franz, G. and Wiener, J.M. (2008). From space syntax to space semantics: A behaviorally and perceptually oriented methodology for the efficient description of the geometry and to-pology of environments. *Environment and Planning B: Planning and Design*, 35(4), 574–592.

Friedman-Hill, E. (2003). *Jess in Action*. Manning Publications.

Fu, X., Guo, D., Liu, X., Pan, G., Qiao, Y., Sun, D. and Steinhardt, N.S. (2002). *Chinese Architecture*. Yale University Press.

Gamma, E., Helm, R., Johnson, R. and Vlissides, J. (2002). Design patterns: Abstraction and reuse of object-oriented design. pp. 701–717. *In*: B. Manfred and D. Ernst (Eds.), *Software Pioneers*. Springer-Verlag New York, Inc.

Gero, J.S. (1990). Design prototypes: A knowledge representation schema for design. *AI Magazine*, 11(4), 26–36.

Gero, J.S. and Tang, H.-H. (1999). Concurrent and retrospective protocols and computer-aided architectural design. pp. 403–410. *In*: Z. Xie and J. Quian (Eds.), Fourth Conference on Computer Aided Architectural Design Research in Asia (CAADRIA1999). Shanghai.

Gero, J.S. and Kannengiesser, U. (2004). The situated function-behaviour-structure framework. *Design Studies*, 25(4), 373–391. https://doi.org/10.1016/j.destud.2003.10.010

Gero, J.S. and Kannengiesser, U. (2014). Commonalities across designing: Empirical results. pp. 285–308. *In*: J.S. Gero (Ed.), *Design Computing and Cognition '12: Proceeding of the Fifth International Conference of Design Computing and Cognition (DCC'12)*.

Goldschmidt, G. and Porter, W.L. (Eds.). (2004). *Design Representation*. Springer-Verlag London.

Gu, K. (2013). *The Private Garden of Jiangnan*. Thsinghua University Press.

Gu, N. and Maher, M.L. (2005). Dynamic designs of virtual worlds using generative design agents. pp. 239–248. *In*: *Proceedings of CAAD Futures 2005*. Springer.

Guo, J. (2014). Application of Depth Map software in spatial structure analysis of Master-of-nets Garden. *Chinese Landscape Architecture 8*, 120–124.

Han, C. (2012). The aesthetics of wandering in the Chinese literati garden. *Studies in the History of Gardens and Designed Landscapes: An International Quarterly*, 32(4), 296–301.

Hanson, J. (1998). *Decoding Homes and Houses*. Cambridge University Press.

Hesselgren, L., Charitou, R. and Dritsas, S. (2007). The Bishopsgate Tower case study. *International Journal of Architectural Computing*, 5(1), 61–81. https://doi.org/10.1260/147807707780912912

Hildebrand, G. (1999). *Origins of Architectural Pleasure*. University of California Press. https://books.google.com.au/books?id=R6MpMs3pz0oC

Hillier, B. (1995). *Space is the Machine*. Cambridge: Cambridge University Press.

Hillier, B. and Hanson, J. (1984). *The Social Logic of Space*. Cambridge University Press.

Hillier, B., Hanson, J. and Graham, H. (1987). Ideas are things, an application of Space Syntax to discovering house genotypes. *Environment and Planning B: Planning and Design*, 14, 363–385.

Hillier, B. and Kali, T. (2006). Space syntax: The language of museum space. pp. 282–301. *In*: M. Sharon (Ed.), A Companion to Museum Studies. Blackwell.

Hunt, J.D. (2012). *A World of Gardens*. Reaktion.

Iordanova, I., Tidafi, T., Guité, M., De Paoli, G. and Lachapelle, J. (2009). Parametric methods of exploration and creativity during architectural design: A case study in the design studio. *In*: 13th International Conference on Computer Aided Architectural Design Futures, Montréal.

Jabi, W., Soe, S., Theobald, P., Aish, R. and Lannon, S. (2017). Enhancing parametric design through non-manifold topology. *Design Studies*, 52, 96–114.

Jackson, P.W. and Messick, S. (1965). The person, the product and the response: Conceptual problems in the assessment of creativity. *Journal of Personality*, 33(3), 309–329.

Jacob, R. and Karn, K. (2003). Eye tracking in human-computer interaction and usability research: Ready to deliver the promises. pp. 573–605. *In*: J. Hyönä, R. Radach and H. Deubel (Eds.), *The Mind's Eye: Cognitive and Applied Aspects of Eye Movement Research*. Oxford.

Jeong, S.K. and Ban, Y.U. (2011). Computational algorithms to evaluate design solutions using Space Syntax. *Computer-Aided Design*, 43(6), 664–676. https://doi.org/10.1016/j.cad.2011.02.011

Jiang, H. (2012). *Understanding Senior Design Students' Product Conceptual Design Activities—A Comparison Between Industrial and Engineering Design Students*. National Universigy of Singapore.

Kan, J.W.T. and Gero, J.S. (2005). Can Entropy Indicate the Richness of Idea Generation in Team Designing? *In*: 10th International Conference on Computer-Aided Architectural Design Research in Asia (CAADRIA 2005), New Delhi, India.

Kan, J.W.T. and Gero, J.S. (2008). Acquiring information from linkography in protocol studies of designing. *Design Studies*, 29(4), 315–337. https://doi.org/10.1016/j.destud.2008.03.001

Kan, J.W.T. and Gero, J.S. (2009). Using the FBS ontology to capture semantic design information in design protocol studies. pp. 213–229. *In*: J. McDonnell and P. Lloyd (Eds.), About: Designing, Analysing Design Meetings. Taylor and Francis.

Kan, J.W.T. and Gero, J.S. (2017). *Quantitative Methods for Studying Design Protocols*. Springer.

Kan, J.W.T. and Gero, J.S. (2013). Studying software design cognition. pp. 61-77. *In*: A.v.d. Hoek and M. Petre (Eds.), Software Designers in Action: A Human-Centric Look at Design Work (1st ed.). Chapman and Hall/CRC, London.

Kaplan, S. (1988). Perception and landscape: Conceptions and misconeptions. *In*: J.L. Nasar (Ed.), Environmental Aesthetics. University of Cambridge.

Kaufman, L. and Richard, W. (1969). Spontaneous fixation tendencies for visual forms. *Perception and Psychophysics*, 5, 85–88.

Keswick, M. (1978). *The Chinese Garden: History, Art and Architecture*. St. Martin's Press.

Keswick, M., Jencks, C. and Hardie, A. (2003). *The Chinese Garden: History, Art and Architecture*. Harvard University Press. http://books.google.com.au/books?id=LYpIPcg9k4cC

Klarqvist, B. (1992). *A Space Syntax Glossary* (Vol. 2). Arkitekturforskning.

Knight, T.W. 2000. *Shape Grammars in Education and Practice. History and Prospects*. http://www.mit.edu/~tknight/IJDC/ (accessed 07 Jul 2020).

Koning, H. and Elzenberg, J. (1981). The language of the prairie: Frank Lloyd Wright's Prairie Houses. *Environment and Planning B: Planning and Design*, 8(3), 295–323.

Kruger, C. and Cross, N. (2006). Solution driven versus problem driven design: Strategies and outcomes. *Design Studies*, 27(5), 527–548. https://doi.org/10.1016/j.destud.2006.01.001

Lahti, H., Seitamaa-Hakkarainen, P. and Hakkarainen, K. (2004). Collaboration patterns in computer supported collaborative designing. *Design Studies*, 25(4), 351–371.

Lee, J., Gu, N. and Williams, A.P. (2014). Parametric design strategies for the generation of creative designs. *International Journal of Architectural Computing*, 12(3), 263–282. https://doi.org/10.1260/1478-0771.12.3.263

Lee, J., Ostwald, M. and Gu, N. (2016). Cognitive challenges for teamwork in design. pp. 55–75. *In*: Richard Tucker (Ed.), Collaboration and Student Engagement in Design Education. https://doi.org/10.4018/978-1-5225-0726-0.ch003"pavilion"

Lee, J.H., Gu, N. and Ostwald, M.J. (2019). Cognitive and linguistic differences in architectural design. *Architectural Science Review*, 62(3), 248–260. https://doi.org/10.1080/00038628.2019.1606777

Li, Z. (2011). Visual perception of traditional garden space in Suzhou, China: A case study with space syntax techniques. *In*: 2011 19th International Conference on Geoinformatics, Shanghai, China. https://doi.org/10.1109/GeoInformatics.2011.5980789

Liang, J., Hu, Y. and Sun, H. (2013). The design evaluation of the green space layout of urban squares based on fractal theory. *Nexus Network Journal*, 15(1), 33–49.

Lorenz, W.E. (2003). Fractals and Fractal Architecture [Thesis]. Vienna University of Technology, Vienna.

Lu, S. (2009). From Syntax to plot: The spatial language of a Chinese garden. *In*: D. Koch, L. Marcus and J. Steen (Eds.), *Proceedings of the 7th International Space Syntax Symposium* (Paper 067), Stockholm: KTH.

Lu, S. (2010). Hidden orders in Chinese gardens: Irregular fractal structure and its generative rules. *Environment and Planning B: Planning and Design*, 37(6), 1076–1094.

Lynch, K. (1960). *The Image of the City*. MA: MIT Press.

Maher, M., Roseman, M., Merrick, K. and Macindoe, O. (2006). DesignWorld: An augmented 3D virtual world for multidisciplinary collaborative design. pp. 133–142. *In*: *CAADRIA 2006: Proceedings of the 11th International Conference on Computer Aided Architectural Design Research in Asia*. Osaka, Japan. http://papers.cumincad.org/cgi-bin/works/Show?caadria2006_133

Mandelbrot, B.B. (1977). *Fractals Form, Chance, and Dimension*. W.H. Freeman and Company.

Mandelbrot, B. (1982). *The Fractal Geometry of Nature*. W.H. Freeman and Company.

Maher, M.L. and Kundu, S. (1993). Adaptive design using a genetic algorithm. IFIP WG5.2 Working Comference on Formal Design Methods.

Maher, M.L. and Poon, J. (1996). Modelling design exploration as co-evolution. *Microcomputers in Civil Engineering*, 11(3), 195–210.

Maher, M.L., Poon, J. and Boulanger, S. (1996). Formalising design exploration as co-evolution. pp. 3-30. *In*: J.S. Gero and F. Sudweeks (Eds.), *Advances in Formal*

Design Methods for CAD: Proceedings of the IFIP WG5.2 Workshop on Formal Design Methods for Computer-Aided Design. June 1995. Springer US. https://doi.org/10.1007/978-0-387-34925-1_1

Maher, M.L. and Tang, H.H. (2003). Co-evolution as a computational and cognitive model of design. *Research in Engineering Design*, 11, 47–63.

Marr, D. (1980). Theory of edge detection. *Proceedings of the Royal Society of London B: Biological Sciences*, 207(1167), 187–217.

Marr, D. (2010). *Vision*. MIT Press.

Meyn, S.P. and Tweedie, R.L. (2009). *Markov Chains And Stochastic Stability*. Cambridge University Press.

Minor, E.S. and Urban, D. (2007). Graph theory as a proxy for spatially explicit population models in conservation planning. *Ecological Applications*, 17, 1771–1782.

Mitchell, W.J. (Ed.) (2003). *Beyond Productivity: Information Technology, Innovation, and Creativity*. National Academies Press.

Moore, C. and Allen, G. (1977). *Dimensions: Space, Shape and Scale in Architecture*. McGraw-Hill.

Nguyen, L. and Shanks, G. (2009). A framework for understanding creativity in requirements engineering. *Information and Software Technology*, 51(3), 655–662. https://doi.org/10.1016/j.infsof.2008.09.002

Nickerson, S.R. (1994). Enhancing creativity. *In*: R.J. Sternberg (Ed.), Handbook of Creativity. Cambridge University Press.

Norman, D. and Draper, S. (1986). *User Centered System Design: New Perspectives on Human-Computer Interaction*. Lawrence Erlbaum Associates.

Nourian, P., Rezvani, S. and Sariyildiz, S. (2013). A syntactic architectural design methodology: Integrating real-time space syntax analysis in a configurative architectural design process. pp. 1–15. *In*: Y. Kim, H.T. Park and K.W. Seo (Eds.), The 9th International Space Syntax Symposium. Seoul, South Korea.

Ostwald, M. (1993). Virtual urban space: Field Theory (Allegorical Textuality) and the search for a new spatial typology. *Transition*, 43(4–24), 64–65.

Ostwald, M. (1997). Virtual urban futures. pp. 125–144. *In*: D. Holmes (Ed.), Virtual Politics: Identity and Community in Cyberspace. Sage Publications.

Ostwald, M. (2011). The mathematics of spatial configuration: Revisiting, revising and critiquing justified plan graph theory. *Nexus Network Journal*, 13(2), 445–470.

Ostwald, M. and Dawes, M. (2013). Differentiating between Line and Point Maps Using Spatial Experience: Considering Richard Neutra's Lovell House. *Nexus Network Journal*, 15(1), 63–81. https://doi.org/10.1007/s00004-012-0134-4

Ostwald, M. and Dawes, M.J. (2018). *The Mathematics of the Modernist Villa: Architectural Analysis Using Space Syntax and Isovists*. Birkhäuser.

Ostwald, M.J. (2013). The fractal analysis of architecture: Calibrating the box counting method using scaling coefficient and grid disposition variables. *Environment and Planning B: Planning and Design*, 40, 644–663.

Ostwald, M.J. and Vaughan, J. (2013a). Limits and errors: Optimising image pre-processing standards for architectural fractal analysis. *Architectural Science Research*, 7, 1–20.

Ostwald, M.J. and Vaughan, J. (2013b). Representing architecture for fractal analysis: A framework for identifying significant lines. *Architectural Science Review*, 56(3), 242–251.

Ostwald, M.J. and Vaughan, J. (2016). The Fractal Dimension of Architecture. Birkhäuser.

Ostwald , M.J., Vaughan, J. and Tucker, C. (2008). Characteristicvisual complexity: Fractal dimensions in the architecture of Frank Lloyd Wright and Le Corbusier. pp. 217–231. *In*: K. Williams (Ed.), Nexus VII: Architecture and Mathematics. Turin: Kim Williams Books.

Peng, Y. (1986). *Analysis of Chinese Classical Garden.* Architecture and Building Press.

Peponis, J. and Wineman, J. (2002). Spatial Structure of Environment and Behavior. *In*: R. Bechtel and A. Churchman (Eds.), Handbook of Environmental Psychology. John Wiley and Sons, Inc.

Phare , D., Gu, N. and Ostwald, M. (2016). Representation in collective design: Are there differences between expert designers and the crowd? *In*: Y. Luo (Ed.), Cooperative Design, Visualization, and Engineering. Cham.

Phare, D.M., Gu, N. and Ostwald, M. (2018). Representation in design communication: Meaning-making in a collective context. *Frontiers in Built Environment*, 4(36). https://doi.org/10.3389/fbuil.2018.00036

Pourmohamadi, M. and Gero, J.S. (2011). LINKOgrapher: An analysis tool to study design protocols based on FBS coding scheme. 294–303. *In*: S. Culley, B. Hicks, T. McAloone, T. Howard and Y. Reich (Eds.), Design Theory and Methodology. pp. Design Society.

Razzouk, R. and Shute, V. (2012). What is design thinking and why is it important? *Review of Educational Research*, 82(3), 330–348. https://doi.org/10.3102/0034654312457429

Rinaldi, B.M. (2011). *The Chinese Garden: Garden Types for Contemporary Landscape Architecture.* Birkhäuser.

Rowe, C. and Slutzky, R. (1963). *Transparency: Literal and Phenomenal.* Vol. 8. The MIT Press on behalf of Perspecta.

Russell, S. and Norvig, P. (2020). *Artificial Intelligence: A Modern Approach.* 4th edition. Prentice Hall.

Salim, F. and Burry, J. (2010). Software openness: Evaluating parameters of parametric modeling tools to support creativity and multidisciplinary design integration. pp.483–497. *In*: D. Taniear et al. (Eds.), *Proceedings of Computational Science and Its Applications (ICCSA 2010).* Springer-Verlag.

Sarkar, P. and Chakrabarti, A. (2011). Assessing design creativity. *Design Studies*, 32(4), 348–383. https://doi.org/10.1016/j.destud.2011.01.002

Schnabel, M.A. (2007). Parametric designing in architecture. pp. 237–250. *In*: A. Dong, A. Vande Moere and J.S. Gero (Eds.), *CAADFutures '07: Proceedings of the Twelfth International Conference on Computer Aided Architectural Design Futures.* Sydney. http://papers.cumincad.org/data/works/att/cf2007_237.content.pdf

Schön, D.A. (1983). *The Reflective Practitioner: How Professionals Think in Action.* Basic Books. http://books.google.com/books?id=ceJIWay4-jgC

Schön, D.A. (1992). Designing as reflective conversation with the materials of a design situation. *Knowledge-Based Systems*, 5(1), 3-14. https://doi.org/10.1016/0950-7051(92)90020-g

Scott, S. (1993). Complexity and Mystery as Predictors of Interior Preferences. *Journal of Interior Design*, 19(1), 25–33.

Simon, H.A. (1969). *The Sciences of the Artificial*. M.I.T. Press.

Simon, H.A. (1973). The structure of ill-structured problems. *Artificial Intelligence*, 4, 81–204.

Singhal, S. and Zyda, M. (1999). *Networked Virtual Environments: Design and Implementation*. ACM Press, New York.

Stamps, A.E. (2002). Fractals, skylines, nature and beauty. *Landscape and Urban Planning*, 60(3), 163–184.

Sternberg, R.J. and Lubart, T.I. (1999). The concept of creativity: Prospects and paradigms. pp. ix, 490 p. *In*: R.J. Sternberg (Ed.), Handbook of Creativity. Cambridge University Press.

Stiny, G. and Gips, J. (1972). Shape grammars and the generative specification of painting and sculpture. pp. 1460–1465. *In*: C.V. Freiman (Ed.), *Proceedings of Information Processing 71*. Amsterdam.

Stiny, G. and Mitchell, W.J. (1978). The Palladian grammar. *Environment and Planning B: Planning and Design*, 5(1), 5–18.

Stiny, G. and Mitchell, W.J. (1980). The grammar of paradise: On the generation of Mughul gardens. *Environment and Planning B: Planning and Design*, 7(2), 209–226.

Sun, P. (2012). The contrast interpretation of Chinese Classical Gardens between space syntax theory and traditional theories—The space research of Chengde Mountain Resort (Doctoral dissertation). Beijing Forestry University.

Tong, J. (1997). *Glimpses of Gardens in Eastern China*. Architecture and Building Press.

Turner, A. (2001). Depthmap: A program to perform visibility graph analysis. 3rd International Space Syntax Symposium, Atlanta.

Ullman, S. (2000). *High-level Vision: Object Recognition and Visual Cognition*. MIT Press.

Vinson, N.G. (1999). Design guidelines for landmarks to support navigation in virtual environments. pp. 278–285. *In*: M.W. Altom and M.G. Williams (Eds.), *Proceedings of the ACM CHI 99 Human Factors in Computing Systems Conference*. ACM Press.

Wang, S. and Wang, B. (2013). Configurational diagram of Chinese traditional architecture: Path network analysis in space syntax. *The Architect*, 162(4), 84–90.

Ware, C. and Mikaelian, H.T. (1987). An evaluation of an eye tracker as a device for computer input. ACM CHI+GI'87 Human Factors Conference, New York.

Weber, R., Choi, Y. and Stark, L. (2002). The impact of formal properties on eye movement during the perception of architecture. *Journal of Architectural Planning and Research*, 19(1), 57–68.

Woodbury, R. (2010). *Elements of Parametric Design*. Routledge. http://books.google.com.au/books?id=HIM3QAAACAAJ

Woodbury, R., Aish, R. and Kilian, A. (2007). Some patterns for parametric modeling. pp. 222–229. *In*: *Proceedings of the 27th Annual Conference of the Association for Computer Aided Design in Architecture*. Halifax, Nova Scotia.

Wooldridge, M. and Jennings, N.R. (1995). Intelligent agents: Theory and practice. *Knowledge Engineering Review*, 10(2), 115–152.

Wortmann, T. and Tunçer, B. (2017). Differentiating parametric design: Digital workflows in contemporary architecture and construction. *Design Studies*, 52, 173–197. https://doi.org/10.1016/j.destud.2017.05.004

Wu, N.I. (1963). *Chinese and Indian Architecture: The City of Man, the Mountain of God, and the Realm of the Immortals*. G. Braziller.

Yu, R. and Gero, J.S. (2017). Exploring designers' cognitive load when viewing digital representations of spaces: A pilot study. *In*: A. Chakrabarti and D. Chakrabarti (Eds.), Research into Design for Communities. Volume 1. *Proceedings of ICoRD 2017*. Springer. https://doi.org/10.1007/978-981-10-3518-0_40

Yu, R., Gero, J.S. and Gu, N. (2013). Impact of using rule algorithms on designers' behavior in a parametric design environment: Preliminary results from a pilot study. *Proceedings of the 15th International Conference on Computer Aided Architectural Design Futures (CAAD FUTURES 2013)*. Shanghai, China.

Yu, R., Gero, J.S. and Gu, N. (2013). Impact of using rule algorithms on designers' behavior in a parametric design environment: Preliminary results from a pilot study. pp. 13–22. *In*: J. Zhang and C. Sun (Eds.), *Global Design and Local Materialization: 15th International Conference on Computer Aided Architectural Design Futures (CAAD FUTURES 2013), Proceedings*. Springer. https://doi.org/10.1007/978-3-642-38974-0_2

Zeiler, W., Savanovic, P. and Quanjel, E. (2007). Design decision support for the conceptual phase of the design process. *Proceeding of International Association of Societies of Design Research (IASDR07)*, Hongkong.

Zhou, W. (1999). *Classical Garden of China*. Tsinghua University Press.

Zhuangzi. (365-286 BC). *Zhuangzi jin zhu jin shi*. Zhonghua Shuju.

Conclusion

At the start of this book, in Chapter 1, we introduced the concept of computational design, and throughout the following chapters we discussed its properties in terms of design technology, cognition and environment. The relationships between these three themes were explored, including the way the design environment encompasses both the technology that supports and enables the design process, and the cognitive behaviours that occur in this process. Chapter 2 built on this foundation to describe emerging computational design technologies in the AEC and design industries, including their history, characteristics, applications and the results of recent research. The chapter also introduced two technologies in detail – parametric and generative design – which are amongst the most recent and influential contemporary computational design approaches. Moving into the second theme, Chapter 3 reviewed current research on design cognition, design thinking and creativity, and discussed the formal research methods used in these fields, including protocol and biometric approaches. Thereafter, this chapter focused on the context of computational design, presenting key findings and theories related to design cognition in specific design environments. Chapter 4 focused on these environments, presenting a series of case studies that apply the theories and methodologies introduced in Chapter 3. The case studies examined the impact of different types of computational design environments on designers and their outputs. Building on these insights and connecting the three themes, this concluding chapter describes the establishment of a conceptual model, demonstrating the relationship between design technology, cognition and environment. Using this structured framework, the chapter concludes with a discussion on future trends in design technology development and their impacts on designers.

5.1. A conceptual model

As described in Chapter 2, design technology has evolved through various stages: from paper-based, using physical models, to computer-aided drafting and modelling, to parametric and generative design, and most recently, to the virtual and augmented world. Throughout this evolutionary process of design technology, designers' cognitive behaviours have been continuously affected and changed. The study of designers' cognition, however, provides the basis and evidence for the development of design technology, as demonstrated in Chapter 3. The purpose of cognitive studies in design is to explore sets of mental processes, strategies and knowledge areas employed by designers whilst designing (Visser, 2004). Both design cognition and technology have an impact on the design environment, being two significant components of it. As we have discussed in Chapter 4, the design environment is made up of an integrated combination of tools, techniques and systems that collectively comprise the ecosystem of the designer. Importantly, the design environment potentially serves more than one designer, and requires a consideration of the communication or interaction that occurs across different types of real and virtual spaces. To explore the relationship between these three elements of computational design, a conceptual model of computation design is developed in the following steps. The first of these steps was described in Chapter 1, and following the content of Chapters 2, 3 and 4, it is now possible to further elaborate the model.

Step 1. In computational design, the Design Environment (D_{Env}) encompasses the technology that supports and enables the design process and the cognitive operations and behaviours that occur in this process (Figure 5.1). As such, D_{Env} is the product of both tools or technological enablers and the thought processes and actions that occur while operating these tools. Design Technology (D_{Tec}) is the set of tools that enables the modelling, visualisation, analysis and generation of design components.

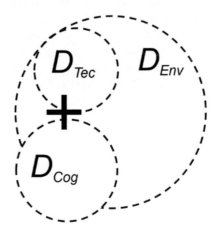

Figure 5.1. Step 1 of the conceptual model of computational design.

Design Cognition (D_{Cog}) is the set of mental processes, behaviours and operations that occur during the design process. These two comprise, but are also not entirely contained within, the boundaries of the design environment. This is because there can be outside influences on each designer's cognitive operations.

Arguably, D_{Env} is greater than the sum of both technology and cognition, as it can include additional factors (systems) that are intrinsic to design operations. These might include quality assurance mechanisms, contractual conditions, documentation and archival systems, all of which are part of the environment but are neither enablers nor cognitive processes. The combination of D_{Cog} and D_{Tec} is core to the design environment, but $D_{Env} > (D_{Tec} + D_{Cog})$.

Step 2. The catalyst for design technology is not necessarily contained within the design environment. External drivers for design technology (E_{Tec}) include advances in computer software, hardware and interfaces. These may, or may not, be motivated by the intention to support design, but they can be adopted or adapted for this purpose. It is important to acknowledge these drivers, although they are not a direct part of the environment supporting the design process. Just as D_{Tec} may be shaped by advances from outside D_{Env}, so can several external factors shape cognition (E_{Cog}) and have a great influence on D_{Env}. These E_{Cog} factors may include education, enculturation and professional or industry related factors (Figure 5.2).

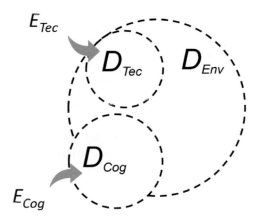

Figure 5.2. Step 2 of the conceptual model of computational design.

Step 3. As mentioned in Chapter 3, design creativity has always been a core issue in design studies. Although creative processes cannot guarantee creative outcomes, common characteristics are found in the design process that can lead to creative outcomes. Creative design processes require particular combinations of problem-finding, idea-generating and problem-solving processes (Osborn, 1963; Parnes, 1981). Design creativity in the design process is affected by design technology, designers' thinking and the design environment. Therefore, creativity in design (D_{Create}) emerges from particular combinations and alignments of cognitive operations and technological enablers in an environment, such that $(E_{Cog} + D_{Cog}) + D_{tec} \times D_{Env}$ leads to D_{Create}. In a sense, the modelling of these factors suggests a convergence, or spiraling centre to the environment, where D_{Create} may emerge (Figure 5.3).

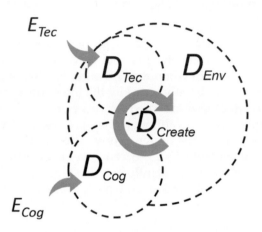

Figure 5.3. Step 3 of the conceptual model of computational design.

Step 4. As stated in Chapter 1, various design technologies support the design process and form the design environment in different ways. While D_{Tec} shapes the design environment, D_{Tec} can also be divided into three categories, some of which may overlap (Figure 5.4).

(i) Foundational tools (D_{Tec1}) are those for generating 2D graphics, 3D models, printed/fabricated media and other basic visualisations that aim to directly support general design processes.

(ii) Supporting tools (D_{Tec2}) are those that directly shape or support the design environment, including collaborative visualisation or modelling tools, virtual world platforms and related simulation and analytical tools.

(iii) Assisting tools (D_{Tec3}) operate at a deeper cognitive level, supporting the search for original, novel, creative or innovative ideas, forms or solutions. These include tools such as those assisting parametric designers in managing generative and parametric versioning.

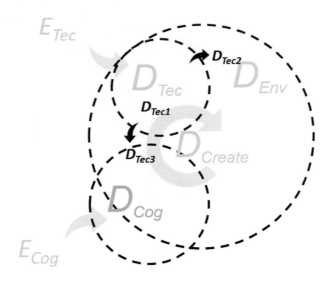

Figure 5.4. Step 4 of the conceptual model of computational design.

Step 5. Some specific design environments entail unique relationships between design cognition and technology. As demonstrated in the previous chapters, technology employed in a computational design process can potentially shape a designer's thought processes and behaviours in both the problem and solution spaces (Chien and Yeh, 2012; Mitchell, 2003). Such problem–solution coevolution processes during design are an important indicator of achieving design creativity. Three types of relationships are identified in this step (Figure 5.5).

(i) Design Cognition 1 (D_{Cog1}) is the process during which designers formulate, in parallel, a problem and idea(s) for a solution. Dorst and Cross (Cross and Cross, 1998; Dorst and Cross, 2001) argue that this co-evolution process is vital for supporting the highest level of creative design. Therefore, this co-evolution of design problem and solution is an indicator of achieving Design Creativity ($E_{Cog} \rightarrow D_{Create}$).

(ii) Design Environment 1 (D_{Env1}) refers to design ideation environments, such as generative and parametric design environments. This type of design ideation environment arises from the use of assisting tools ($D_{Tec3} \rightarrow D_{Env1}$) and it has a significant impact on cognition (D_{Cog}) in comparison with other design environments whose primary focus is to support design documentation or presentation.

(iii) Design Environment 2 (D_{Env2}) refers to collaborative virtual environments. It arises from the use of supporting tools ($D_{Tec2} \rightarrow D_{Env2}$) and it also has a significant impact on design cognition (D_{Cog}).

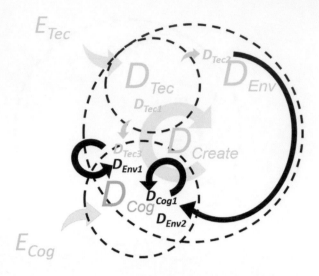

Figure 5.5. Step 5 of the conceptual model of computational design.

When all the operations in the model are combined (Figure 5.6), the intricate relationships amongst design environment, technology and cognition can be visualised and tested. This conceptual model highlights and abstracts these three important aspects of computational design. Table 5.1 lists the definitions of each component within the conceptual model.

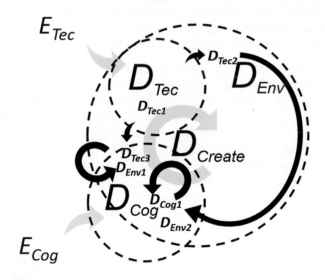

Figure 5.6. The complete conceptual model of computational design based on design environment, design technology and design cognition.

Table 5.1. Components of the conceptual model.

Component	Abbreviation	Definition
Design Environment	D_{Env}	Design Environment encompasses the technology that supports and enables the design process and the cognitive operations and behaviours that occur in this process.
Design Technology	D_{Tec}	Design Technology is the set of tools that enables the modelling, visualisation, analysis and generation of design components.
Design Cognition	D_{Cog}	Design Cognition is the set of mental processes, behaviours and operations that occur during the design process.
External Drivers for Design Technology	E_{Tec}	External Drivers for Design Technology include advances in computer software, hardware and interfaces.
External Cognition	E_{Cog}	External Cognition factors are influences from outside the design environment that shape the cognitive processes occurring within. These may include education, enculturation and profession related factors.
Design Creativity	D_{Create}	Design Creativity refers to creative processes and/or creative outcomes expressed during the design process.
Design Technology 1	D_{Tec1}	Design Technology 1 refers to 'foundational tools' including those for generating 2D graphics, 3D models, printed/fabricated media and other basic visualizations.
Design Technology 2	D_{Tec2}	Design Technology 2 refers to 'supporting tools' which directly shape the design environment, including collaborative visualization or modelling tools, virtual world platforms and related simulation and analytical tools.
Design Technology 3	D_{Tec3}	Design Technology 3 includes 'assisting tools' which enable operations at the deeper cognitive level. These support the search for original or novel, creative and innovative ideas, forms or solutions.

(Contd.)

Table 5.1. (*Contd.*)

Component	Abbreviation	Definition
Design Cognition 1	D_{Cog1}	Design Cognition 1 is a process during which designers formulate a problem and idea(s) for a solution in parallel.
Design Environment 1	D_{Env1}	Design Environment 1 refers to design ideation environments, such as generative and parametric design environments, arising from D_{Tec3}.
Design Environment 2	D_{Env2}	Design Environment 2 refers to collaborative virtual environments arising from D_{Tec2}.

5.2. Looking into the future of computational design

The conceptual model of computational design developed above provides a formal structure to further explore computational design. It can be used to guide designers in computational design practice through better understanding design technology, cognition and environment, and to guide scholars in critically reviewing current work and planning future research. Using this framework, the following sections discuss the implications and future developments of computational design – considering technology, cognition and environment as a whole – to better support design and designers.

5.2.1. Design technology: Implications and future developments

As discussed in Chapter 2, computational design has promoted a paradigm shift for designers. It has not only made the automated generation, modification and optimisation of design possible, it has become possible to create immersive experiences, to see or even feel a design in ways that were never possible or imaginable in the past. These have been collectively achieved by way of the three categories of design technologies (D_{Tec1}, D_{Tec2}, D_{Tec3}).

D_{Tec3} supports the search for creative ideas or solutions. This category includes parametric and generative design tools that are not only able to produce complex free-form buildings (using parameters and variables), but are able to assist designers with analysing and optimising the generated design through establishing and manipulating rule algorithms, too. This allows designers to effectively consider a wide range of design issues and rapidly test a large number of variations. In recent years, plug-in tools that can be embedded into major generative and parametric software are emerging to serve various design purposes, such as apps for *Grasshopper* (www.food4rhino.com). Recent computational tools based on generative approaches, such as *Archistar* (archistar.ai) and *Giraffe* (www.giraffe. build), can be applied to broader domains ranging from urban design to property development, and the generative algorithms are directed with more refined and complex parameters including real-time data sets about site factors, cost and sustainability. It is anticipated that this trend will continue with the development of future generative and parametric tools that are able to assist designers to explore new possibilities, especially those with a high-level complexity that requires effective optimisation across a range of social, environmental and cultural concerns.

With the advancement of computational design, more realistic and engaging digital design representations will continue to remain important as they enable designers to more readily explore the feasibility of different design solutions in relation to real contexts, and to more effectively communicate and engage with stakeholders and end users. This technology relies on basic D_{Tec1} tools for 2D and 3D modelling and interactive visualisation. It is expanded in BIM, where 3D models (building-related information modelling) have added 4D (time factors), 5D analytical simulations (Eastman, 2008) and even nD for realistic (both visual and functional) simulations. Further development of those computational design tools will enable more advanced digital design representations to assist effective design thinking and communication (Self et al., 2014).

D_{Tec2} are tools that directly support the design environment. They include collaborative, virtual and analytical systems and technology. For example, many virtual worlds and game engines have enabled designers to collaborate remotely without being physically co-located. Virtual Reality (VR) developments have proven to be effective, providing realistic simulated environments where designers can explore a range of possibilities (Wang and Dunston, 2013). Augmented Reality (AR) developments, on the other hand, can enhance a designer's perceptions by providing design environments that complement both the real and physical worlds (Morrison et al., 2011). Further developments of D_{Tec2} are especially helpful for the purposes of enhancing communication in a collaborative design setting (Goldschmidt and Porter, 2004), with support for distant collaboration and more seamless integration of virtuality and reality.

5.2.2. Design cognition: Implications and future developments

The technologies used by designers have changed, and are continuing to change the way designers think and act. Design cognition research has explored designers' thought processes and their connection to various design issues, such as behaviour, creativity and the co-evolution process (D_{Cog1}).

Traditionally, the research methods used for studies in design cognition have included protocol analysis and observations of, or interviews with, designers. Arising from developments in neuroscience, multiple new methods have recently become available in this field. For example, biometric approaches such as eye tracking, EEG and FRMI can be used to diversify and refine research results. These approaches can reveal a designer's cognitive behaviours and thinking patterns by collecting and exploring biometric responses. It is anticipated that design neuroscience will be one of the important future directions in cognitive design studies.

With the wider adoption and recognition of computational design in contemporary practices, design expertise is not only about the designer's specialist domain knowledge, but also about their capabilities in applying computational tools and methods. This is especially evident in design firms that have adopted parametric and generative design environments. Over time, computational designers have become familiar with geometrical modelling and parametric rule-based operations, developing so-called "design patterns" which they use repetitively (Yu and Gero, 2015). Such patterns, which can be seen as an "induction" process whereby the designer generalises samples from their own design experience, or from observation of other designers, abstract the problem-solution pairs for different design situations (Fowler, 2003). These patterns are then formalised for re-use in future designs. From this growing practice we can infer that when designers use programming code or scripts in their design process, they also exhibit similar characteristics of developing and using patterns appropriate to the given environment. This is evidently a typical process in generative and parametric design environments, and although it is not well understood, it will remain important, especially with the continuing evolution of new design technologies.

5.2.3. Design environment: Implications and future developments

The design environment, or design ecosystem, is made up of an integrated combination of tools, techniques and systems. This understanding emphasises the importance of understanding future developments in computational design technology, cognition and environments in an integrated manner. For example, particular future developments of design tools may include hardware, such as VR glasses, software such as next generation parametric design applications, and new collaborative platforms. It is important, however, to note that these technological advancements alone do not guarantee a sound or creative design environment. The effect of the design environment is closely related to design cognition, which directly considers and measures the impacts on designers. Therefore, as highlighted in our conceptual model, the effective developments of both D_{Env1} and D_{Env2} and their positive influences (on design and designers) should be critically informed by cognitive design studies focusing on computational design tools. Further research is needed in this regard to not only develop sound cognitive design research in these areas, but also to establish feasible business models for implementing technology developments in practice.

5.3. Conclusion

This book has examined three important elements in computational design – design technology, cognition and environment – and explored the dynamic relationships between them. Using a combination of critical reviews, experimental results and case studies, it has developed a conceptual model of computational design, elaborating a series of complex relationships and connections.

In the last decade, the field of design technology has advanced and evolved rapidly, from computer aided drafting to 3D modelling, and then to generative and parametric design. The features and capabilities of these tools have also been enhanced to accommodate virtual, augmented and mixed realities, data analytics, social computing and citizen sciences. These sets of emerging computational design tools provide novel approaches that can assist designers in generating, optimising and communicating design. The integration and adoption of new design technology is also changing the way both designers work and the design industries operate. Designers' capabilities in applying computational tools and methods in design are, therefore, becoming an important sub-set of design knowledge. To effectively utilise these new technologies, to provide sound design environments that support designer decision making will, require evidence drawn from design cognition. Current cognitive design studies adopt multiple methodologies, such as protocol analysis and biometric methods, to provide formal and rigorous approaches to understanding designers' cognitive behaviours and their thought processes in design. The sustainable future development of computational design will require an integrated, well balanced and informed approach considering all three elements.

References

Chien, S.-F. and Yeh, Y.-T. (2012). On creativity and parametric design—A preliminary study of designer's behaviour when employing parametric design tools. *In*: H. Achten, J. Pavlíčk and J. Huhín (Eds.), *Proceedings of 30th International Conference on Education and Research in Computer Aided Architectural Design in Europe (eCAADe 2012)*. pp. 245–253. Czech Republic.http://papers. cumincad.org/cgi-bin/works/Show?ecaade2012_223

Cross, N. and Cross, C. (1998). Expertise in engineering design. *Research in Engineering Design*, 10, 141–149.

Dorst, K. and Cross, N. (2001). Creativity in the design process: Co-evolution of problem-solution. *Design Studies*, 22(5), 425–437. https://doi.org/10.1016/s0142-694x(01)00009-6

Eastman, C.M. (2008). *BIM handbook: A guide to building information modeling for owners, managers, designers, engineers and contractors*. Wiley. http://books. google.com.au/books?id=IioygN0nYzMC

Fowler, M. (2003). *Patterns of enterprise application architecture*. Addison-Wesley. http://books.google.com.au/books?id=FyWZt5DdvFkC

Goldschmidt, G. and Porter, W.L. (Eds.) (2004). *Design Representation*. Springer-Verlag.

Mitchell, W.J. (Ed.) (2003). *Beyond Productivity: Information Technology, Innovation, and Creativity*. National Academies Press.

Morrison, A., Mulloni, A., Lemmelä, S., Oulasvirta, A., Jacucci, G., Peltonen, P., Schmalstieg, D. and Regenbrecht, H. (2011). Collaborative use of mobile augmented reality with paper maps. *Computers & Graphics*, 35(4), 789–799.

Osborn, A.F. (1963). *Applied Imagination: Principles and Procedures of Creative Problem-solving*. Scribner.

Parnes, S.J. (1981). *The Magic of your Mind*. Bearly Limited.

Self, J., Lee, S.-g. and Bang, H. (2014). *Understanding the Complexities of Design Representation*. *In*: Proceedings of 2013 Ancient Futures: Design and/or Happiness. Asian Digital Art & Design Association & Korean Society of Design Science, Korea.

Visser, W. (2004). Dynamic aspects of design cognition: Elements for a cognitive model of design. [Research report]. RR-5144, INRIA. https://hal.inria.fr/inria-00071439/document

Wang, X. and Dunston, P. (2013). Tangible mixed reality for remote design review: A study understanding user perception and acceptance. *Visualization in Engineering*, 1(1), 1–15.

Yu, R. and Gero, J. (2015). Design patterns from empirical studies in computer-aided design. pp. 493–506. *In*: C. Gabriela, D. Moreno and J. Moara (Eds.), Computer-aided Architectural Design Futures—The Next City—New Technologies and the Future of the Built Environment. Springer.

Appendix: Coding Example for Case Study 1

Designer 6

PDE session

ID	Timespan	Content	1st Coding	2nd Coding	Final Coding
1	0:00.0 - 0:12.3	Well, I will start from rhino and grasshopper task, as you can see here, is the site.	R-K	R-K	R-K
2	0:12.3 - 0:28.3	The definition I usually used is the division of the sun	F-K	F-K	F-K
3	0:28.3 - 0:35.6	There are some definitions by depth and height, normally I use, a pack, but I don't know if I will be quick with this.	Be-R	Be-K	Be-R
4	0:48.5 - 0:59.3	So, I start with parameters of time zone, latitude, longitude, and I have a season – a summer icon for the calculation on this.	Be-R	Be-K	Be-R
5	1:08.1 - 1:16.5	The site, I want to create is a pyrography... something like this	Be-K	S-K	Be-K
6	1:16.5 - 1:26.4	So first, I will start from the rooms, class rooms, and meeting rooms	F-K	F-K	F-K

7	1:26.4 - 1:34.1	If we think a little bit about the street here, we have	F-K	F-K	F-K
8	1:34.1 - 1:38.4	This point, maybe the maxim connectivity point go	Be-K	Be-K	Be-K
9	1:38.4 - 1:44.2	We have 5 streets on this point, so	Bs-K	Bs-K	Bs-K
10	1:44.2 - 1:50.6	I think this is correct for the entrance.	F-K	Be-K	F-K
11	1:50.6 - 1:59.1	Since we have the park here,	Be-K	F-K	F-K
12		I think it will be nice to put another entry here	Be-K	Be-K	Be-K
13	1:59.1 - 2:08.1	I think I will put parking in this zone,	F-K	Be-K	Be-K
14	2:08.1 - 2:10.3	Because it is a secondary road	F-K	F-K	F-K
15	2:10.3 - 2:16.6	So I will put a parking here,	F-K	Be-K	Be-K
16		Something like this (draw a rectangle)	S-K	S-K	S-K
17	2:16.6 - 2:26.1	So I will start with a meeting room.	F-K	F-K	F-K
18		For example (draw a circle)	S-K	S-K	S-K
19	2:26.1 - 2:33.4	Something like that (re-draw the circle)	S-K	S-K	S-K
20	2:33.4 - 2:38.6	Then I will put classroom over here	F-K	F-K	F-K
21		(Draw a circle)	S-K	S-K	S-K
22	2:38.6 - 2:50.4	And then a tutorial room	F-K	F-K	F-K
23		like this, at this point.	S-K	S-K	S-K
24	2:50.4 - 2:56.5	Here I made an ellipse not a circle because I	S-K	S-K	S-K
25	2:56.5 - 3:04.4	Think it is necessary to create a façade within the parking	F-K	Be-K	F-K
26	3:04.4 - 3:12.4	I will make this a bit smaller	Be-K	S-K	S-K

27		(Delete and re-draw a circle)	S-K	S-K	S-K
28	3:12.4 - 3:21.1	So we have to connect those	Be-K	Be-K	Be-K
29		(Draw curve to connect circles)	S-K	S-K	S-K
30	3:21.1 - 3:35.1	I will try to find its centre	Be-K	S-K	Be-K
31		(Draw to find the centre of the triangle)	S-K	S-K	S-K
32	3:35.1 - 3:47.8	And we will connect those spaces	Be-K	Be-K	Be-K
33	3:47.8 - 3:56.2	(Draw curves)	S-K	S-K	S-K
34	3:56.2 - 4:01.5	Three elements (delete the triangle)	S-K	S-K	S-K
35	4:01.5 - 4:13.8	Then we can offset, I think 1.5 m (offset)	S-K	S-K	S-K
36	4:13.8 - 4:17.4	(Delete previous curve)	S-K	S-K	S-K
37	4:17.4 - 4:26.0	So I will split all of these	Be-K	Be-K	Be-K
38		(Split)	S-K	S-K	S-K
39	4:26.8 - 4:36.0	To get an organic group to work, operate	Be-K	Be-K	Be-K
40	4:36.0 - 4:49.5	Then we can make..., something like that	Bs-K	Be-K	Be-K
41		(Fillet the curves)	S-K	S-K	S-K
42	4:49.5 - 4:57.0	We have this, as the first group. we join curves	S-K	S-K	S-K
43	4:57.0 - 5:08.0	So I will make this place a contour group	Be-K	S-K	S-K
44	5:08.0 - 5:14.4	I will take this group into grasshopper (set "cur" component)	Be-R	S-R	S-R
45	5:14.4 - 5:26.1	I will explode this on site (set "explode" component)	Be-R	S-R	Be-R
46	5:26.1 - 5:31.7	(Set "list item")	Be-R	Be-R	Be-R
47	5:31.7 - 5:50.0	(Change properties of "list item")	S-R	S-R	S-R
48		I will copy it to create the contour lines	Be-R	F-K	F-K

49	5:50.0 - 5:57.6	I will create a loft, simple... to operate	Be-R	S-K	S-K
50	5:57.6 - 6:13.4	(Set "move")	S-K	S-K	S-K
51	6:13.4 - 6:18.2	I think I need to reverse it (set "x" unit, and "reverse")	Be-R	Be-R	Be-R
52	6:18.2 - 6:33.7	I will use I series, start with 5 (set "series" component)	Be-R	Be-R	Be-R
53	6:33.7 - 6:43.3	Each 5 metres, and how to make steps I need	Be-K	Be-K	Be-K
54	6:43.3 - 6:46.0	I'll put a slider (set parameters) here	S-R	S-R	S-R
55	6:46.0 - 6:50.7	(Set constraints)	Be-R	Be-R	Be-R
56	6:50.7 - 6:57.1	(Change parameters)	S-R	S-R	S-R
57	6:57.1 - 7:04.9	Then I will intersect those lines (set "cct" component)	Be-R	Be-R	Be-R
58	7:04.9 - 7:11.0	(Connect component)	S-R	S-R	S-R
59	7:11.1 - 7:16.0	Select the intersection points and create the final lines	S-K	S-K	S-K
60		Set "list item")	Be-R	Be-R	Be-R
61	7:16.0 - 7:25.7	This is starting points, and end points(set "list item")	Be-R	Be-R	Be-R
62		(And set parameters)	S-R	S-R	S-R
63	7:25.7 - 7:31.2	(Set "line" component)	S-K	S-K	S-K
64	7:31.2 - 7:35.7	"Blind preview"	Bs-K	Bs-R	N
65	7:35.7 - 7:41.8	I will intersect these new lines with a create line	Be-K	Be-R	Be-K
66	7:41.8 - 7:50.0	Select this (set curve)	S-K	S-K	S-K
67	7:50.0 - 7:58.4	Intersect one more time (set "ccx")	Be-R	Be-R	Be-R
68	7:58.4 - 8:00.0	Ok, so we have intersection points, so...	Bs-R	Bs-R	Bs-R
69	8:00.0 - 8:05.4	What we need to create now is a vertical line that	Be-R	S-K	S-K

70	8:05.4 - 8:25.4	Gives me a force weight, I think it will be 2 or 3 metres in the first and it will increase at the mid points.	Be-R	Be-R	Be-R
71	8:25.4 - 8:31.7	(Set "line" component).	S-K	S-K	S-K
72	8:31.7 - 8:35.1	Set the direction (set "z" direction).	Be-R	Be-R	Be-R
73	8:35.1 - 8:41.6	So we try to meet the points we have	Be-R	Be-R	Be-R
74		(Change parameters)	S-R	S-R	S-R
75	8:41.6 - 8:49.0	I want to create a point in the middle of each line.	Be-K	Be-R	Be-K
76	8:49.0 - 8:54.6	So first we need to create the lines	S-K	S-K	S-K
77	8:54.6 - 9:01.0	(Set "list item")	Be-R	Be-R	Be-R
78	9:01.0 - 9:09.7	(Change parameters)	S-R	S-R	S-R
79	9:09.7 - 9:16.4	(Set "line" component)	S-K	S-K	S-K
80	9:16.4 - 9:20.2	Yes, perfect	Bs-K	Bs-K	Bs-K
81	9:20.2 - 9:23.4	I will try to get these inter-medium points and create a line,	Be-R	Be-K	Be-K
82	9:23.4 - 9:25.9	(Set parameters)	S-R	S-R	S-R
83	9:25.9 - 9:33.1	I think it is perfect.	Bs-K	Bs-K	Bs-K
84		(Change parameters)	S-R	S-R	S-R
85	9:33.1 - 9:50.5	(Set "list item")	Be-R	Be-R	Be-R
86		Check data	Bs-R	Bs-R	Bs-R
87	9:50.5 - 9:58.3	We will try different numbers (copy "list item" and	Be-R	Be-R	Be-R
88		"Line" component)	S-K	S-K	S-K
89	9:58.3 - 10:05.4	(Change parameters)	S-R	S-R	S-R
90	10:05.4 - 10:08.5	This is nice	Bs-K	Bs-K	Bs-K
91	10:08.5 - 10:17.0	But now those points are not on the same side (rotate the model)	Bs-K	Bs-K	Bs-K

92	10:17.0 - 10:29.0	So now what I want to do is to create lines between all those lines	Be-K	Be-K	Be-K
93	10:29.0 - 10:35.8	The contour line will finally give me the surface	Be-K	F-K	Be-K
94	10:35.8 - 11:04.2	(Rotate the model)	Bs-K	Bs-K	Bs-K
95		This is the problem grasshoppers always have, and I prefer to solve it with a rhino.	N	N	N
96	11:04.2 - 11:47.3	So now I will bake these lines, including points (bake).	N	Bs-K	N
97	11:47.3 - 12:02.9	So I want to bring it here to get the mid points to make all the plants	Be-K	F-K	Be-K
98	12:02.9 - 12:16.3	(Draw line)	S-K	S-K	S-K
99	12:16.3 - 12:25.5	So I will make it a bit easier (draw line)	S-K	S-K	S-K
100	12:25.5 - 12:37.5	So I will create these guide lines	Be-K	Be-K	Be-K
101	12:37.5 - 12:44.8	When I have all the lines, I will create a loft	Be-K	S-K	S-K
102	12:44.8 - 12:48.4	So I will create a topography to work on	Be-K	F-K	Be-K
103	12:48.4 - 12:53.0	And this topography to create the façade of the building	F-K	F-K	F-K
104	12:52.9 - 14:45.8	(Draw lines)	S-K	S-K	S-K
105	14:45.8 - 15:01.2	So let's start making this, I connect starting point, mid-point and end point. (connecting points)	S-K	Be-R	Be-R
106	15:01.2 - 18:46.0	The grasshopper may be easier, but for this one, it needs a test, so I prefer a rhino	N	N	N
107		(Connecting points)	S-K	S-K	S-K

108	18:46.0 - 18:51.4	Now we have all lines, so I will create a loft	Be-K	S-K	S-K
109		(Rotate the model)	Bs-K	Bs-K	Bs-K
110	18:51.4 - 19:23.0	I will create a surface using loft to get my topography	Be-K	F-K	Be-K
111		(Loft)	S-K	S-K	S-K
112	19:23.0 - 19:29.6	The façade is based on this	Be-K	Bs-K	Bs-K
113		(Rotate the model)	Bs-K	Bs-K	Bs-K
114	19:29.6 - 19:39.4	(Set "surface" component)	S-K	S-K	S-K
115	19:39.4 - 19:54.4	I will use "..box to triangulate this surface (set "tri—area" component)	Be-R	Be-R	Be-R
116	19:54.4 - 20:07.6	Because, to design a building, it is the way to transmit .	N	Be-K	Be-K
117	20:07.6 - 20:13.1	(Set parameters)	S-R	S-R	S-R
118		Now I will divide it as 50	Be-K	Be-K	Be-K
119		I think it will be too much		Bs-K	Bs-K
120	20:13.1 - 20:20.6	(Change parameters)	S-R	S-R	S-R
121	20:20.6 - 20:24.1	(Change parameters)	S-R	S-R	S-R
122	20:24.2 - 20:28.7	Surface, 25 (connect sliders).	S-R	S-R	S-R
123		Here we have flow.. (rotate the model)	Bs-K	Bs-K	Bs-K
124	20:37.6 - 20:55.6	(Connect components)	S-R	S-R	S-R
125	20:54.2 - 21:09.2	(Checking problem and re-connecting)	Bs-R	Bs-R	Bs-R
126	21:09.2 - 21:18.7	So we have here the sun radiation levels (rotate the model)	Bs-K	Bs-K	Bs-K

127	21:18.7 - 21:31.6	What we want to do is to search the minimal radiation and panels for windows	Be-K	F-K	Be-K
128	21:34.8 - 21:55.0	For the other part, I will put some green panels or something like this to cool what is over it.	Be-K	Be-K	Be-K
129	21:55.0 - 22:00.1	(Rotate the model)	Bs-K	Bs-K	Bs-K
130		Those are the entrances	F-K	F-K	F-K
131	22:00.1 - 22:05.6	(Rotate the model)	Bs-K	Bs-K	Bs-K
132	22:05.6 - 22:12.1	We'll see... here this is the data (check data)	Bs-R	Bs-R	Bs-R
133	22:12.1 - 22:40.1	100, 1000 (check previous script data)	Bs-R	Bs-R	Bs-R
134	22:40.1 - 22:48.3	I want to divide this domain into three parts, I think.	Be-R	Be-R	Be-R
135	22:48.3 - 22:52.6	Or just into one for the windows	F-K	Be-R	Be-R
136	22:52.6 - 22:56.5	And the other for the green panels	F-K	Be-R	Be-R
137	22:56.5 - 23:01.6	But I think it will be too much (rotate the model)	Bs-K	Bs-K	Bs-K
138	23:01.6 - 23:06.8	Too many panels with window	Bs-K	Bs-K	Bs-K
139	23:06.8 - 23:09.0	So I don't like this (rotate the model)	Bs-K	Bs-K	Bs-K
140	23:13.0 - 23:29.1	If I make this. (hide the model)	N	N	N
141	23:29.1 - 23:37.6	Here are the lines of the traffic	F-K	F-K	F-K
142	23:37.6 - 23:48.0	So I think it will be good to create another one	F-K	Be-K	Be-K
143	23:48.0 - 24:02.3	And I want to go to the park this way (unhide a curve)	Be-K	Be-K	Be-K
144	24:02.3 - 24:33.1	Let's say, from the street, intersect (draw traffic line)	Be-K	Be-K	Be-K

145	24:33.1 - 24:46.2	Another one is from the parking (draw another traffic line)	Be-K	Be-K	Be-K
146		Right now I think it is more towards completion	Bs-K	Bs-K	Bs-K
147	24:46.2 - 24:53.5	So now I want to select all the panels in the zone of these lines, and create a different panel	S-K	S-K	S-K
148	25:03.2 - 25:09.8	So let's make an offset of three metres (offset curves)	S-K	S-K	S-K
149	25:28.5 - 25:35.9	Intersect, split (split curves)(split and trim)	S-K	S-K	S-K
150	26:16.3 - 26:19.7	So let's make a 5 metre street	F-K	F-K	F-K
151	26:19.7 - 26:41.2	(Trim and fillet)	S-K	S-K	S-K
152		So we have a really nice place	Bs-K	Bs-K	Bs-K
153	26:41.2 - 26:48.7	(Trim and fillet)	F-K	S-K	S-K
154		Ok, then we have the park		F-K	F-K
155	26:48.7 - 27:09.9	Let me join , all these lines	S-K	S-K	S-K
156	27:09.9 - 27:21.1	Now I will close all these lines to create a region (close the lines)	Be-K	S-K	Be-K
157	27:21.1 - 27:33.5	I will select all the panels inside the region, depart form the others to create these panels with rules, as I said (close the lines)	Be-R	Be-R	Be-R
158	27:53.1 - 28:07.8	(Set "curve" component)	S-K	S-K	S-K
159	28:07.8 - 28:17.9	I'll find a region (set "different solid " components)	S-K	S-K	S-K
160	28:17.9 - 28:33.8	So cool, I have one.. (set "curve" and connect component)	S-K	S-R	S-K

161	28:33.8 - 28:50.7	(Set "xy" plane)	Be-R	Be-R	Be-R
162	28:50.7 - 29:03.0	The region created, perfect (set "planar" component)	S-K	S-K	S-K
163	29:03.0 - 29:09.0	So here is the region	F-K	Bs-K	Bs-K
164	29:09.0 - 29:22.6	I have all the panels here to make it look well	Be-K	Bs-K	Bs-K
165	29:22.6 - 29:33.3	We have the centre, (set "area")	Be-R	Be-R	Be-R
166	29:33.3 - 29:41.7	(Rotate the model)	Bs-K	Bs-K	Bs-K
167	29:41.7 - 30:05.2	Right now, I want to create surfaces, one surface is one from here to the end of the park, it's like creating a bigger park.	F-K	F-K	F-K
168	30:05.2 - 30:13.6	Do not delete this street, but make it a transition	F-K	F-K	F-K
169	30:13.6 - 30:20.2	I've finished this part with lots of surface	Bs-K	S-K	S-K
170	30:20.2 - 30:33.6	So this is one (make a surface in the corner)	S-K	S-K	S-K
171	30:33.6 - 31:03.4	And the other one would be (make surfaces)	S-K	S-K	S-K
172	31:03.4 - 31:10.5	This is big (measure the distance)	Bs-K	Be-K	Bs-K
173	31:10.5 - 31:20.9	(Make surfaces) ok. (set "surface")	S-K	S-K	S-K
174	31:32.2 - 31:48.2	So make the same triangulation at those surfaces	Be-K	S-K	Be-K
175	31:48.2 - 32:02.3	(Unhide component preview)	N	N	N
176	32:02.3 - 32:26.9	This happens, you have to slip the surface and you have to wait for the grasshopper take the original surface, but rhino I don't know why (rotate the model).	N	N	N

177		(Rotate the model)	Bs-K	Bs-K	Bs-K
178	32:26.9 - 32:48.2	I will make this surface (make surfaces)	S-K	S-K	S-K
179	32:48.2 - 33:04.6	(Rotate the model)	Bs-K	Bs-K	Bs-K
180	33:04.6 - 33:35.9	(Re-make the surface)	S-K	S-K	S-K
181	33:35.9 - 34:01.4	I will delete the other part (rotate the model)	S-K	Bs-K	S-K
182	34:01.4 - 34:12.4	(Rotate the model)	Bs-K	Bs-K	Bs-K
183		You know, sometimes, rhino makes these kinds of things.	N	N	N
184	34:12.5 - 34:17.0	We have our topography here	N	F-K	F-K
185	34:17.0 - 34:27.3	Lets get the centres (set "area" component)	Be-R	Be-R	Be-R
186	34:27.3 - 34:35.4	Lets put all the centre together (set "point" component)	S-R	Be-R	S-R
187	35:01.6 - 35:08.4	(Un-preview)	N	N	N
188	35:08.4 - 35:18.3	So lets see which are the points inside these regions (set "contain" component)	Be-R	Be-R	Be-R
189	35:30.4 - 35:36.4	(Check the data)	Bs-R	Bs-R	Bs-R
190	35:36.4 - 35:50.3	So lets see which are equal to one (set "lager" component)	Be-R	Be-R	Be-R
191	35:50.3 - 36:00.0	(Check the data)	Bs-R	Bs-R	Bs-R
192	36:00.0 - 36:11.9	(Change parameters)	S-R	S-R	S-R
193	36:11.9 - 36:29.6	Lets make the surface (set "srf" component)	S-K	S-K	S-K
194	36:29.6 - 36:36.2	(Un-preview)	N	N	N

195	36:36.2 - 36:50.7	(Check previous data)	Bs-R	Bs-R	Bs-R
196	36:50.7 - 36:57.4	I will create a cone of this region (set "cone" component)	S-K	S-K	S-K
197	37:07.5 - 37:13.7	(Set "z" direction)	Be-R	Be-R	Be-R
198		(Set parameters)	S-R	S-R	S-R
199	37:13.7 - 37:26.0	(Set "planar srf" component)	S-K	S-K	S-K
200	37:26.0 - 37:45.7	(Set "include" component)	Be-R	Be-R	Be-R
201	37:45.7 - 38:06.6	The grasshopper is "thinking", I think right now we have 10 minutes to finish	N		N
202	38:06.6 - 38:09.4	(Check previous data)	Bs-R	Bs-R	Bs-R
203	38:09.4 - 38:15.5	(Connected)	S-R	S-R	S-R
204	38:15.5 - 38:30.8	(Un-preview)	N	N	N
205	38:30.8 - 38:42.7	(Unhide)	N	N	N
206	38:42.7 - 38:48.0	(Generate surface using selected points)	S-K	S-K	S-K
207	38:48.0 - 39:02.2	Now there are some points under the surface, they are not inside the region	Bs-R	Bs-K	Bs-K
208	39:02.2 - 39:07.7	So we need to make transform it	Be-R	Be-R	Be-R
209	39:07.7 - 39:11.9	Move (set "move" component)	S-K	S-K	S-K
210	39:11.9 - 39:18.6	We need to reverse it (set "z" unit)	Be-R	Be-R	Be-R
211	39:18.6 - 39:25.7	2 metres (set parameter)	S-R	S-R	S-R
212	39:25.7 - 39:36.1	Now it works well (connect components)	S-R	S-R	S-R

213	39:36.1 - 39:42.4	Waiting	N	N	N
214	39:42.4 - 39:57.0	Now we have all these panels (rotate the model)	Bs-K	Bs-K	Bs-K
215	39:57.0 - 40:05.9	Now do the same to the others	Be-R	S-K	S-K
216	40:05.9 - 40:18.7	I think I will put all of these surface into this (set "srf" component)	Be-R	S-R	S-R
217	40:18.7 - 40:25.3	Because those panels will create the path	F-K	F-K	F-K
218	40:25.3 - 40:34.5	So I will make this variation	Be-R	Be-R	Be-R
219	40:34.5 - 40:39.0	(Check previous data)	Bs-R	Bs-R	Bs-R
220	40:39.0 - 41:07.6	(Connect components)	S-R	S-R	S-R
221	41:07.6 - 41:26.8	Perfect (preview)	Bs-K	Bs-K	Bs-K
222	41:26.8 - 41:42.3	So let's average all of these source together, into two lists	Be-R	Be-R	Be-R
223	41:42.3 - 42:08.5	(Set "average" component)	Be-R	Be-R	Be-R
224	42:08.5 - 42:13.5	Larger, or smaller (set "larger" component)	Be-R	Be-R	Be-R
225	42:13.5 - 42:24.0	(Connect components)	S-R	S-R	S-R
226	42:24.0 - 42:25.8	And then smaller (set "smaller" component)	Be-R	Be-R	Be-R
227	42:25.8 - 42:28.6	(Connect components)	S-R	S-R	S-R
228	42:28.6 - 42:33.5	I dispatch one more time (set "dispatch" component)	Be-R	Be-R	Be-R
229	42:40.3 - 42:52.1	(Connect components)	S-R	S-R	S-R
230	42:52.1 - 43:00.5	Take true values (set "srf" component)	S-K	Be-R	S-K
231	43:00.5 - 43:05.5	(Unhide)	N	N	N

232	43:05.5 - 43:13.5		S-R	S-R	S-R
233	43:13.5 - 43:27.7	Ok, perfectly done. we have all these that will be windows, the other will be grass or green fonts	F-K	F-K	F-K
234	43:27.7 - 43:41.8	So I will take these lines (set "edges" component)	S-K	S-K	S-K
235	43:41.8 - 43:50.8	Now I offset them to create the window	S-K	S-K	S-K
236	43:50.8 - 43:54.4	(Set "joint" component)	S-R	S-K	S-R
237	43:54.4 - 44:00.3	(Set "offset" component)	S-K	S-K	S-K
238	44:00.3 - 44:02.7	Perfect	Bs-K	Bs-K	Bs-K
239	44:02.7 - 44:08.8	No, I need to flip	Be-R	Be-R	Be-R
240		(Rotate the model)	Bs-K	Bs-K	Bs-K
241	44:08.8 - 44:11.1	The line is in the same direction	Bs-R	Bs-R	Bs-R
242	44:11.1 - 44:22.8	(Set "flip")	Be-R	Be-R	Be-R
243	44:22.8 - 44:24.2	(Set "list item")	Be-R	Be-R	Be-R
244	44:24.2 - 44:35.7	(Flatten and connect)	S-R	S-R	S-R
245	44:35.7 - 44:42.5	(Examine the model)	Bs-K	Bs-R	Bs-K
246	44:42.5 - 44:46.0	Ok, perfect	Bs-K	Bs-K	Bs-K
247	44:46.0 - 44:58.3	So I think I would fillet it together with the wall	S-K	S-K	S-K
248	44:58.3 - 45:04.2	So 0.2 (set parameters)	S-R	S-R	S-R
249	45:04.2 - 45:09.1	Ok, perfect	S-K	Bs-K	Bs-K
250	45:09.1 - 45:17.1	Those lines are the glass of the windows	F-K	F-K	F-K
251		(Rotate the model)	Bs-K	Bs-K	Bs-K

252	45:17.1 - 45:26.4	Surface (set "planar" component)	S-K	S-K	S-K
253	45:26.4 - 45:30.1	Perfect (rotate the model)	Bs-K	Bs-K	Bs-K
254	45:30.1 - 45:44.3	We need to create the region between the external line and these lines one more time. (set "region difference" component)	Be-R	Be-R	Be-R
255	45:53.7 - 46:07.7	(Connect components)	S-R	S-R	S-R
256	46:07.7 - 46:11.3	Surface (set "planar" component)	S-K	S-K	S-K
257	46:11.3 - 46:14.8	Perfect	Bs-K	Bs-K	Bs-K
258	46:14.8 - 46:24.1	(Check previous script)	Bs-R	Bs-R	Bs-R
259	46:24.1 - 46:29.8	Ok, perfect	Bs-K	Bs-K	Bs-K
260	46:29.8 - 46:36.0	This is the window	F-K	F-K	F-K
261		(Rotate the model)	Bs-K	Bs-K	Bs-K
262	46:36.0 - 46:51.5	And here we have path panels, and there are ribbons	F-K	F-K	F-K
263	46:51.5 - 46:55.6	So let's bake layers (make layers)(bake)	N	N	N
264	47:38.6 - 47:42.6	(Examine the model) the same	Bs-K	Bs-K	Bs-K
265	47:42.6 - 48:02.3	(Bake)	N	N	N
266	48:02.3 - 48:14.6	So right now we have all of them	Bs-K	Bs-K	Bs-K
267	48:14.6 - 48:37.6	(Hide layers) I am looking for lines, I see it	Be-K	N	N
268	48:37.6 - 48:48.9	(Delete lines outside site boundary)	S-K	S-K	S-K
269	48:48.9 - 48:54.8	(Examine model)	Bs-K	Bs-K	Bs-K

270	48:54.8 - 49:18.5	(Delete lines outside site boundary)	S-K	S-K	S-K
271		Anyway, I will leave this part right now, but I want to show you that this will be the building	N	N	N
272	49:28.1 - 49:38.4	I will make this blue (change colour of the layer)	N	S-K	S-K
273	49:58.3 - 50:03.0	So it will be something like that right now (rotate the model)	Bs-K	Bs-K	Bs-K
274	50:03.0 - 50:05.8	I will delete that part I made, then I think it will be better	S-K	S-K	S-K
275	50:19.8 - 50:22.5	A little bit of work	N	N	N
276	50:22.5 - 50:44.4	But I think this is an interesting test in 40 mins.	N	N	N

GME session

ID	Timespan	Content	1st coding	2nd coding	Final coding
1	0:00.0 - 0:06.2	Well, task 2. Rhino task	R-K	N	R-K
2	0:06.7 - 0:26.2	Let's start with planning. I will create something really classic, I think.	Be-K	Be-K	Be-K
3	0:26.2 - 0:34.0	(Offset site boundary)	S-K	S-K	S-K
4	0:34.0 - 1:01.7	(Draw curves) ok	S-K	S-K	S-K
5	1:01.7 - 1:10.6	Let's start, these are my first lines to make a plan.	N	N	N
6	1:10.6 - 1:14.9	Let's connect these (trim curves)	S-K	S-K	S-K
7	1:14.9 - 1:21.7	So I think this part would be the entrance	F-K	F-K	F-K
8	1:21.7 - 1:31.1	And this part would become something like "built points" to get a different view from this park	F-K	Be-K	Be-K

9	1:35.6 - 1:50.8	So the other part here,is, the sun will be from the north	Be-K	Be-K	Be-K
10	1:50.8 - 1:59.5	So I will create a blind façade in this part	S-K	S-K	S-K
11	1:59.5 - 2:11.5	Another... I will get a window façade here	S-K	S-K	S-K
12	2:11.5 - 2:13.2	To get all the sun this side	Be-K	Be-K	Be-K
13	2:13.2 - 2:24.3	So I think I will put a stairway here, to get	F-K	F-K	F-K
14	2:24.3 - 2:30.9	The classroom on the first floor	F-K	F-K	F-K
15	2:30.9 - 2:38.0	And the ground floor have the general space and open space here	F-K	F-K	F-K
16	2:38.0 - 3:12.9	So this is something like this, in the middle, or... (draw curves)	S-K	S-K	S-K
17	3:12.9 - 3:26.3	Ok, I will split this, and I will rotate it through 90 degrees, -90 degrees (rotate the curve) I will make it from this point, –90 (rotate the curve) ok	S-K	S-K	S-K
18	3:53.3 - 4:00.6	So I will move to the intersection (move points to the intersection)	S-K	S-K	S-K
19	4:00.6 - 4:04.6	Perfect	Be-K	Bs-K	Bs-K
20		(Delete one curve)	S-K	S-K	S-K
21	4:04.6 - 4:14.5	And I will have this, here, this way	Bs-K	Bs-K	Bs-K
22		(Draw curves)	S-K	S-K	S-K
23	4:21.4 - 4:23.4	So here we will have a stairway,	F-K	F-K	F-K
24	4:23.4 - 4:28.2	Upto the first floor, here we will have a path to the general space on the ground floor	F-K	F-K	F-K
25	4:33.4 - 4:47.5	So let's see this, all half here (draw curves)	Bs-K	S-K	S-K
26	4:47.5 - 4:58.3	Let's make it (draw curves)	S-K	S-K	S-K

27	4:58.3 - 5:17.0	Offset this to create a classroom 2 meters long	F-K	Be-K	F-K
28	5:17.0 - 5:19.1	(Offset curves) no	S-K	S-K	S-K
29	5:19.1 - 5:34.1	Let's copy, to create a classroom, like these (copy curves)	S-K	S-K	S-K
30	5:34.1 - 5:37.2	All we have here are two classrooms	Bs-K	Bs-K	Bs-K
31	5:37.2 - 5:44.2	But the first one needs to be created	Be-K	N	Be-K
32	5:44.2 - 5:51.1	Now let's consider the pass,	F-K	F-K	F-K
33		And here is the stairway which leads to the classroom	Be-K	Be-K	Be-K
34	5:51.1 - 5:53.5	And create the points of view here	Be-K	Be-K	Be-K
35	5:53.5 - 6:03.1	Here we have double..	Bs-K	Bs-K	Bs-K
36		You can look at this from ground floor	Be-K	Be-K	Be-K
37	6:03.3 - 6:17.2	So let's make something like this (draw curves)	S-K	S-K	S-K
38	6:17.2 - 6:19.8	Too much	Bs-K	Bs-K	Bs-K
39	6:19.8 - 6:25.3	(Redraw the curve)	S-K	S-K	S-K
40	6:25.3 - 6:30.9	Here we will have a point of view and two classrooms	Bs-K	F-K	Bs-K
41	6:30.9 - 6:57.9	I am just trying to create a simple line to form this stairway, I think here (draw lines)	S-K	S-K	S-K
42	6:57.9 - 7:01.4	Something like this (delete curves)	S-K	S-K	S-K
43	7:01.4 - 7:15.8	So let's move (move curves)	S-K	S-K	S-K
44	7:15.8 - 7:24.1	Ok, and here are the final lines (draw curves)	S-K	S-K	S-K
45	7:24.1 - 7:33.9	But as the sun comes by this way, from the north,	Be-K	Be-K	Be-K

46	7:33.9 - 7:57.1	Let's create something different here, ok, something like this to give the sun entry	Be-K	Be-K	Be-K
47		(Draw curves)		S-K	S-K
48	7:57.1 - 8:01.7	This way and have a different façade to this park	S-K	Be-K	Bs-K
49	8:05.0 - 8:11.9	And have a different plan for different parts	Be-K	Be-K	Be-K
50	8:11.9 - 8:18.5	(Adjust curves)	S-K	S-K	S-K
51	8:18.5 - 8:26.7	See we have this, stairway.	Bs-K	Bs-K	Bs-K
52	8:26.7 - 8:32.0	Move this point (adjust the curve)	S-K	S-K	S-K
53	8:32.0 - 8:38.0	(Delete the curve)	S-K	S-K	S-K
54	8:38.0 - 8:42.5	(Draw curves)	S-K	S-K	S-K
55	8:42.5 - 9:03.2	I think this wall or this façade will be concrete, so	S-K	S-K	S-K
56	9:03.2 - 9:15.7	I will create two languages on concrete and the rest of the building	Be-K	Be-K	Be-K
57	9:15.7 - 9:22.1	So this will be wood and glass, and	S-K	S-K	S-K
58	9:22.1 - 9:28.2	There will be whole concrete here and little openings.	S-K	S-K	S-K
59	9:28.2 - 9:39.3	(Adjust the curve)	S-K	S-K	S-K
60		So let's say this is the main wall	F-K	S-K	S-K
61	9:39.3 - 9:56.9	3.1 meters, say 1 meter (offset curves and close the wall boundary)	S-K	S-K	S-K
62	9:56.9 - 10:05.8	Yes, I want to have a wall with a great entity as a.. with some openings	Be-K	Be-K	Be-K
63	10:11.8 - 10:20.2	Something like equal openings something like that, in this way	Be-K	Be-K	Be-K

64	10:20.2 - 10:26.7	Let's go to this (change to perspective view, rotate the model)	Bs-K	Bs-K	Bs-K
65	10:26.7 - 10:32.6	This line may not be here	Bs-K	Bs-K	Bs-K
66	10:32.6 - 10:33.7	(Delete curves)	S-K	S-K	S-K
67	10:33.7 - 10:36.1	Again, this will be the stairway	F-K	F-K	F-K
68	10:36.1 - 10:39.8	And the space here to look	F-K	F-K	F-K
69	10:39.9 - 10:45.4	(Rotate the model)	Bs-K	Bs-K	Bs-K
70	10:45.4 - 10:57.6	This is something like 5 meters, or 4 meters, 5 meter (draw a line in 3d view)	S-K	S-K	S-K
71	10:57.6 - 11:12.4	5 meters, no, we need more than 5 meters, doubled, 10 meters (draw a line in 3d view)	S-K	S-K	S-K
72	11:12.4 - 11:17.1	For two floors, two stage	Be-K	Be-K	Be-K
73	11:17.1 - 11:29.1	Let's copy this (copy curves)	S-K	S-K	S-K
74	11:28.4 - 11:34.8	So something like this (rotate the model)	Bs-K	Bs-K	Bs-K
75	11:34.8 - 11:53.1	I think it will be nice if we make some lines as a façade	Be-K	S-K	S-K
76	11:53.1 - 12:01.2	(Draw a line) something like that	S-K	S-K	S-K
77	12:01.2 - 12:07.4	(Rotate the model)	Bs-K	Bs-K	Bs-K
78	12:07.4 - 12:17.8	And then here I have,, 12 meters, (draw curve on the right view)	S-K	S-K	S-K
79	12:17.8 - 12:24.0	(Copy curves)	S-K	S-K	S-K
80	12:24.0 - 12:33.0	So I have these faces	Bs-K	Bs-K	Bs-K
81		(Make surface)	S-K	S-K	S-K

82	12:33.0 - 12:47.8	Let's copy this (copy curves)	S-K	S-K	S-K
83	12:47.8 - 13:01.1	We can start, the stairway here, and then (make a surface)	Be-K	S-K	S-K
84	13:01.1 - 13:08.4	We have this face anyway (make surface)	S-K	S-K	S-K
85	13:08.4 - 13:15.0	(Rotate the model) ok, that's fine, but	Bs-K	Bs-K	Bs-K
86	13:15.0 - 13:21.2	We'll move this (move curve)	S-K	S-K	S-K
87	13:21.2 - 13:23.2	(Delete surface)	S-K	S-K	S-K
88	13:23.2 - 13:39.6	I am making this façade or surface looking to combine them together, simply form (making surface)	S-K	S-K	S-K
89	13:39.6 - 13:50.9	Nothing difficult,	Bs-K	N	N
90		Just have a nice form (rotate the model)		Bs-K	Bs-K
91	13:50.9 - 13:54.0	(Making surface)	S-K	S-K	S-K
92	13:54.0 - 13:57.8	So, let's see (rotate the model)	Bs-K	Bs-K	Bs-K
93	13:57.8 - 14:10.0	No, I want,... anyway this is not a simple plane, it's complicated, I don't want it to be	N	N	N
94	14:10.0 - 14:15.9	But what I want is something like this,	Be-K	Be-K	Be-K
95	14:15.9 - 14:22.0	So I will remake it(move the surface)	S-K	S-K	S-K
96	14:25.1 - 14:33.0	Yes, something like.. (rotate the model)	Bs-K	Bs-K	Bs-K
97	14:33.0 - 14:35.5	(Delete surface)	S-K	S-K	S-K
98	14:35.5 - 14:40.4	Let's put here a minim...	Be-K	Be-K	Be-K
99	14:40.4 - 14:48.4	Ok, let's say, this is 5 (change length of curve)	S-K	S-K	S-K

100	14:48.4 - 14:53.4	This is 5, too (change length of curve)	S-K	S-K	S-K
101	14:53.4 - 14:57.7	(Rotate the model)	Bs-K	Bs-K	Bs-K
102	14:57.7 - 15:01.7	Then we have here 10, 10 meters.	S-K	S-K	S-K
103	15:01.9 - 15:07.3	This would not be here (delete curves)	S-K	S-K	S-K
104	15:07.3 - 15:21.6	I think I will put things here, this and this (copy curves)	S-K	S-K	S-K
105	15:21.6 - 15:30.9	(Rotate the model) let's see	Bs-K	Bs-K	Bs-K
106	15:30.9 - 15:42.2	No, could be... (copy curves)	S-K	S-K	S-K
107	15:42.2 - 15:47.9	Because I want some part of these to be the points to be opened	Be-K	Be-K	Be-K
108	15:47.9 - 15:56.8	No, let's put it on tense... (create and delete curves)	S-K	S-K	S-K
109	15:56.8 - 16:09.4	(Copy curves)	S-K	S-K	S-K
110	16:09.4 - 16:17.8	So let's see, we have this (make surfaces)	S-K	S-K	S-K
111	16:17.8 - 16:19.8	(Rotate the model)	Bs-K	Bs-K	Bs-K
112	16:19.8 - 16:25.4	And this (make surfaces)	S-K	S-K	S-K
113	16:33.6 - 16:55.1	(Rotate the model) I am not convinced	Bs-K	Bs-K	Bs-K
114	16:55.1 - 16:59.4	So let's see, while I will get another panel (make surfaces)	S-K	S-K	S-K
115	16:59.4 - 17:02.2	So here is a wall	Bs-K	S-K	S-K
116	17:02.2 - 17:05.6	(Make surfaces)	S-K	S-K	S-K
117	17:05.6 - 17:12.9	(Rotate the model)	Bs-K	Bs-K	Bs-K

118	17:12.9 - 17:29.6	Well, I don't know if offsetting this line will be easier than extending this edge, because if I extend it, I have to morph this point	Be-K	Be-K	Be-K
119	17:37.1 - 17:45.1	Anyway, I will try, surface, extend	S-K	S-K	S-K
120	17:45.1 - 17:51.5	5 meter, may not be so much (extend edge)	S-K	S-K	S-K
121	17:51.5 - 18:04.5	I will decompose this surface (decompose surface)	S-K	S-K	S-K
122	18:04.5 - 18:15.5	Ok, perfectly (rotate the model)	Bs-K	Bs-K	Bs-K
123	18:15.5 - 18:37.5	I will create lift.. (create ground surface edge)	S-K	S-K	S-K
124	18:37.5 - 18:40.9	Lift 5 meters (copy edge)	S-K	S-K	S-K
125	18:40.9 - 18:46.8	Make it (make surface)	S-K	S-K	S-K
126	18:46.8 - 18:50.2	This way, perfect (rotate the model)	Bs-K	Bs-K	Bs-K
127	18:50.2 - 18:57.4	So extrude, 3.3 (extrude the surface)	S-K	S-K	S-K
128	18:57.4 - 19:01.0	(Rotate the model)	Bs-K	Bs-K	Bs-K
129	19:01.0 - 19:26.8	And then put this here, the stairway (draw curves and make a surface)	S-K	S-K	S-K
130	19:26.8 - 19:30.2	Here we have the stairway	Bs-K	Bs-K	Bs-K
131	19:30.2 - 19:37.9	So let's join this line (join curves)	Be-K	S-K	Be-K
132	19:37.9 - 19:46.4	Divide, to create some points, like 100 (divide the curve)	S-K	S-K	S-K
133	19:46.4 - 19:59.2	I put a little stick here in the wall, to have this wall in wood and glass	Be-K	S-K	S-K
134		(Rotate the model)	Bs-K	Bs-K	Bs-K

135	19:59.2 - 20:05.4	It will be really easy in grasshopper, but	N	N	N
136	20:05.4 - 20:17.0	3.3, ok, this is what I want	Bs-K	Bs-K	Bs-K
137		(Make a rectangle)	S-K	S-K	S-K
138	20:17.0 - 20:21.2	Move it (move the rectangle)	S-K	S-K	S-K
139	20:21.2 - 20:27.7	Copy (copy the rectangle)	S-K	S-K	S-K
140	20:27.7 - 20:30.1	No (delete the rectangle)	S-K	S-K	S-K
141	20:30.1 - 21:33.3	Copy, it's going to be a lot	N	Be-K	N
142		(Copy the rectangle)	S-K	S-K	S-K
143	21:33.3 - 21:41.1	Ok. I have this copy (copy)	S-K	S-K	S-K
144	21:41.1 - 21:48.4	Need to rotate, this, and this (rotate the rectangle)	S-K	S-K	S-K
145	21:48.4 - 22:15.9	Copy this (copy)	S-K	S-K	S-K
146	22:15.9 - 22:21.7	Ok, so let's create a layer called "wood" (create new layer)	N	N	N
147	22:21.7 - 22:53.1	I forgot this (copy)	S-K	S-K	S-K
148	22:53.1 - 23:02.3	Extrude, solid, stick (Extrude)	S-K	S-K	S-K
149	23:02.3 - 23:18.2	Select all these sticks, let's split it with this plane (split)	S-K	S-K	S-K
150	23:18.2 - 23:56.8	Perfect	Bs-K	Bs-K	Bs-K
151		(Delete the sticks upper)	S-K	S-K	S-K
152	23:56.7 - 24:02.5	Let's make another layer, that is concrete (make new layer)	N	N	N
153	24:02.5 - 24:13.9	(Rotate the model)	Bs-K	Bs-K	Bs-K
154	24:13.9 - 24:21.9	(Hide layers, change properties of layers)	N	N	N

155	24:21.9 - 24:27.0	(Rotate the model)	Bs-K	Bs-K	Bs-K
156	24:27.0 - 24:38.1	Select all these (select curves) Maybe exactly the same	N	S-K	Be-K
157	24:49.9 - 25:00.6	Solid perfectly (extrude)	S-K	S-K	S-K
158	25:00.6 - 25:16.9	(Rotate the model)	Bs-K	Bs-K	Bs-K
159		Split (split)	S-K	S-K	S-K
160	25:16.9 - 25:22.1	(Turn off the layers)	N	N	N
161	25:22.1 - 25:55.3	Delete (delete stick upper)	S-K	S-K	S-K
162	25:55.3 - 26:09.7	(Move points to shorten the sticks)	S-K	S-K	S-K
163	26:09.7 - 26:15.5	Perfect (rotate the model)	Bs-K	Bs-K	Bs-K
164	26:15.5 - 26:21.6	So I will need to make the same with this line	Be-K	Be-K	Be-K
165	26:21.6 - 26:40.5	I want to get 1 meter here, to get to the roof, so	Be-K	Be-K	Be-K
166	26:40.5 - 26:45.4	Scissors (trim curve)	S-K	S-K	S-K
167	26:45.4 - 26:51.0	Divide (divide curve)	S-K	S-K	S-K
168	26:51.0 - 27:14.7	Analyse distance between points, 1.3 (check distance)	Bs-K	Bs-K	Bs-K
169	27:14.7 - 27:24.9	Divide, length, 1.3. perfect (divide curve)	S-K	S-K	S-K
170	27:24.9 - 28:59.0	(Copy rectangle)	S-K	S-K	S-K
171	28:59.0 - 29:15.9	So I am copying this point here, and moving it, 1.2 at least(copy and move points)	S-K	S-K	S-K
172	29:15.9 - 29:22.9	Something like this, I want all the things to get here (rotate the model) to get all these grid windows	Bs-K	Bs-K	Be-K

173	29:26.8 - 30:00.1	So (select all the rectangles) perfect	Bs-K	Bs-K	Bs-K
174	30:00.1 - 30:06.5	Yes extrude here (extrude the rectangles)	S-K	S-K	S-K
175	30:06.5 - 30:29.6	And we have it, offset this, 1 meter? (offset)	S-K	S-K	S-K
176	30:29.6 - 30:57.7	Connect those lines, perfect (trim curves)	S-K	S-K	S-K
177	30:57.7 - 31:19.5	Let's make a new layer of this, called window or glass (make new layers)	N	N	N
178	31:19.5 - 31:27.5	Extrude the line, let's get to the high point (extrude)	S-K	S-K	S-K
179	31:27.5 - 31:32.2	Split it	S-K	S-K	S-K
180	31:32.2 - 31:42.0	This is the problem (rotate the model) (delete the surface)	Bs-K	Bs-K	Bs-K
181	31:42.0 - 31:49.4	I want to extend this edge to this line	S-K	Be-K	Be-K
182	31:49.4 - 31:58.0	This would be opened, and this will be the view point, something is happening here	Be-K	Be-K	Be-K
183		(Rotate the model)		Bs-K	Bs-K
184	31:58.0 - 32:02.3	(Copy sticks)	S-K	S-K	S-K
185	32:02.3 - 32:05.2	Ok (rotate the model)	Bs-K	Bs-K	Bs-K
186	32:05.2 - 32:17.5	And then extend the surface (extend surface), let's say 30.	S-K	S-K	S-K
187	32:17.5 - 32:23.7	1 meter (extend the curve)	S-K	S-K	S-K
188	32:23.7 - 32:56.6	So let's take all together, split it (split the wall)	S-K	S-K	S-K
189	32:56.6 - 33:06.3	(Delete extra curves)	S-K	S-K	S-K
190	33:06.3 - 33:15.5	Let's extrude those surfaces, 3.3 (extrude surface)	S-K	S-K	S-K

191	33:15.5 - 33:22.9	(Rotate the model) perfect	Bs-K	Bs-K	Bs-K
192	33:22.9 - 33:27.3	And this will be 1 meter (extrude)	S-K	S-K	S-K
193	33:27.3 - 33:55.5	Ok, well, I need to split this (split surface)	S-K	S-K	S-K
194	33:55.5 - 33:59.6	(Make surface)	S-K	S-K	S-K
195	33:59.6 - 34:04.7	Ok, perfect (rotate the model)	Bs-K	Bs-K	Bs-K
196	34:04.7 - 34:12.2	And here, I will need to create a concrete with some openings which are different	Be-K	S-K	S-K
197	34:17.3 - 34:28.0	Let's see in front (hide some layers)	Bs-K	Bs-K	Bs-K
198	34:28.0 - 35:38.7	So, let's see, this is a kind of opening (draw windows in the front view)	S-K	S-K	S-K
199	35:38.7 - 35:43.3	(Evaluate the façade)	Bs-K	S-K	Bs-K
200	35:43.3 - 35:57.0	Later there will be more time for working on this kind of opening, but the concept is there	Be-K	N	N
201	35:57.0 - 36:28.0	(Draw openings)	S-K	S-K	S-K
202	36:28.0 - 37:11.2	Take all these lines, and I will extrude them, solid, and then split them, to create openings (select curves)	Be-K	Be-K	Be-K
203	37:11.2 - 37:18.7	Ok (rotate the model)	Bs-K	Bs-K	Bs-K
204	37:18.0 - 37:23.6	(Extrude curves)	S-K	S-K	S-K
205	37:23.6 - 37:30.3	They are solid? yes, they are solid (check the model)	Bs-K	Bs-K	Bs-K
206	37:30.3 - 37:32.8	I don't want solid (delete the extrude curves)	S-K	S-K	S-K

207	37:32.9 - 37:38.0	(Extrude again)	S-K	S-K	S-K
208	37:38.0 - 37:49.1	So (rotate the model)	Bs-K	Bs-K	Bs-K
209	37:49.1 - 38:16.7	(Change properties of layer)	N	N	N
210	38:16.7 - 38:29.7	I need to extrude this (extrude)	S-K	S-K	S-K
211	38:29.7 - 38:33.5	Perfect (rotate the model)	Bs-K	Bs-K	Bs-K
212	38:33.5 - 39:00.1	Split these openings (split)	S-K	S-K	S-K
213	39:00.1 - 39:02.4	Reject this (delete extruded curves)	S-K	S-K	S-K
214	39:02.4 - 39:22.9	So the first on the face will be the light and the second one will pass the glass layer (check the openings on the wall)	Be-K	Bs-K	Bs-K
215	39:22.9 - 40:21.2	(Delete the surface on the openings)	S-K	S-K	S-K
216	40:21.2 - 40:31.1	(Check the location of the surface)	Bs-K	Bs-K	Bs-K
217	40:31.1 - 41:02.5	(Delete the surface on the openings)	S-K	S-K	S-K
218	41:02.5 - 41:05.0	Ok (rotate the model)	Bs-K	Bs-K	Bs-K
219	41:05.0 - 41:51.0	Then pass those to the last layer, glass (select the surface and change to the layer)	Be-K	N	N
220	41:51.0 - 42:16.8	Delete the glass (change to the other layer)	S-K	S-K	S-K
221	42:16.8 - 42:21.5	Ok (unhide the layers)	N	N	N
222	42:21.5 - 42:29.5	So this will be the building (rotate the model)	Bs-K	Bs-K	Bs-K
223	42:29.5 - 42:35.1	Big one (rotate the model)	Bs-K	Bs-K	Bs-K

224	42:35.1 - 42:40.7	This will be wood, concrete, and these would be view points	S-K	Bs-K	S-K
225	42:40.7 - 42:49.0	Interesting	Bs-K	Bs-K	Bs-K
226	42:49.0 - 42:55.4	And then they will have two classroom here and another space too (rotate the model)	F-K	F-K	F-K
227	42:55.4 - 42:58.6	I am just thinking where is the parking (rotate the model)	F-K	F-K	F-K
228	42:58.6 - 43:03.8	We have all on the entrance, here is the parking (draw rectangle)	F-K	F-K	F-K
229	43:03.8 - 43:08.2	(Rotate the model) yes, perfect	Bs-K	Bs-K	Bs-K
230	43:08.2 - 43:25.9	This point, and we have a really nice façade here (rotate the model)	Bs-K	Bs-K	Bs-K

Appendix of Images Sources

Figure 2.2, source from: Gu, N., Yu, R. and Behbahani, P.A. (2018). Parametric design: Theoretical development and algorithmic foundation for design generation in architecture. pp. 1–22. *In:* B. Sriraman (Ed.). Handbook of the Mathematics of the Arts and Sciences. Springer.

Figure 3.2, source from: Yu, R., Gero, J. and Gu, N. (2013). Impact of using rule algorithms on designers' behaviour in a parametric design environment: Preliminary result from a pilot study. pp. 13–22. *In:* J. Zhang and C. Sun (Eds.). Global Design and Local Materialization. 15th International Conference, CAAD Futures 2013. Springer.

Figure 4.2, Figure 4.5 and Figure 4.6, source from: Yu, R., Ostwald, M. and Gu, N. (2015). Empirical evidence of designers' cognitive behaviour in a parametric design environment and geometricmodelling environment. pp. 2437–2446. *In:* V. Popovic, A. Blackler, D.-B. Luh, N. Nimkulrat, B. Kraal and Y. Nagai (Eds.). Interplay: Proceedings of the 6th International Congress of International Association of Societies of Design Research (IASDR 2015). Brisbane, Australia.

Figure 4.4, Figure 4.7, Figure 4.12, Figure 4.13, Figure 4.14, source from: Yu, R. (2014). Exploring the Impact of Rule Algorithms on Designers' Cognitive Behaviour in a Parametric Design Environment. PhD Thesis, University of Newcastle, Australia.

Figures 4.8–4.11, source from: Yu, R., Gero, J. and Gu, N. (2015). Architects' cognitive behaviour in parametric design. *International Journal of Architectural Computing (IJAC)*, 13(01), 83–102.

Figures 4.15–4.19, source from: Yu, R., Gu, N., Ostwald, M. and Gero, J. (2015). Empirical support for problem-solution co-evolution in a parametric design environment. *Artificial Intelligence for Engineering Design, Analysis, and Manufacturing (AIEDAM)*.

Figures 4.21–4.28, source from: Yu, R. and Gero, J. (2018). Using eye-tracking to study designers' cognitive behaviour when designing with CAAD. pp. 443–452. *In:* P. Rajagopalan and M.M. Andamon (Eds.).

Engaging Architectural Challenges of Higher Density. 52nd International Conference of the Architectural Science Association (ASA) 2018. Melbourne, Australia.

Figures 4.29–4.35, source from: Gu, N. and Maher, M.L. (2014). *Designing Adaptive Virtual Worlds*. De Gruyter.

Figures 4.37–4.39 and images in Table 4.14–4.16, source from: Yu, R., Ostwald, M. and Gu, N. (2015). Parametrically generating new instances of traditional Chinese Private Gardens that replicate selected socio-spatial and aesthetic properties. *Nexus Network Journal: Architecture and Mathematics*, 17(3), 807–829.

Figures 4.40–4.44, source from: Yu, R., Gu, N. and Ostwald, M. (2016). The mathematics of spatial transparency and mystery: Using syntactical data to visualise and analyse the properties of the Yuyuan Garden. *Visualisation in Engineering*.

Images in Tables 4.16–4.18, source from: Yu, R., Ostwald, M. and Gu, N. (2015). Parametrically generating new instances of traditional Chinese Private Gardens that replicate selected socio-spatial and aesthetic properties. *Nexus Network Journal: Architecture and Mathematics*, 17(3).

Images in Table 4.19, source from: Yu, R., Gu, N. and Ostwald, M. (2018). Evaluating creativity in parametric design environments and geometric modelling environments. *Architectural Science Review*, 61(6), 443–453.

Index

3D Studio, 19
3M company, 73

A

AAEmtech, 32
ACADIA, 33
Active Worlds, 128, 145
AR-media, 42, 43
Archeoguide, 43, 44
ArchiCAD, 42, 127
ArchImage, 157
ARKit, 43
ARQuake, 43
ARtGlass, 43
ASCAAD, 33
AUGmentecture, 42
Augmented Reality (AR), 15, 35-37, 48, 205
AutoCAD, 16, 18-20
Autodesk, 16, 18-20, 42
AutoFlix, 19
AutoShade, 19

B

Baroque, 46
Bezier curves, 13, 17
BIM, 16, 18, 20, 32, 39, 42, 126, 205
Blender3D, 19
Building Information Modelling (BIM), 16, 32

C

CAAD Futures, 33

CAADRIA, 33
CAD, 3, 5, 7, 11-21, 27, 39, 42, 43, 45, 48, 59, 68, 77-79, 102, 104, 127, 131, 132, 185
Cathode Ray Tube, 12
CATIA, 18, 19, 31
Cellular automata, 21, 23, 25
Church of Colònia Güell, 31
Computer-aided design, 11, 68, 132
Connectivity, 103, 147-149, 151-157, 164, 166-168, 186, 210
Control, 18, 27, 28, 31-33, 37, 44, 45, 80, 82, 141, 145, 148, 165, 167, 169, 170
Convex map, 164, 168

D

Daoist, 166
DesignWorld, 128
Dessault Systèmes, 18
Digital Project, 31
DirectX, 18
Drift Magnitude, 165, 167, 172, 173
Dynamo, 20

E

eCAADe, 33
Ecotect, 32
Electronic Drafting Machine, 12
EnTiTi, 42
ETABS, 32, 34

F

Federation Square, 33

Folfogram, 42
Foster+Partners, 32
Fractal dimension, 103, 157, 162
Frank Gehry, 31

G

Gaudi, 31
Gaze Point, 73
Gehry, Frank, 31
Gehry Technologies, 31
GenerativeComponents, 32, 35
Generative design, 4, 7, 20-26, 28, 33,
    48, 59, 68, 102, 103, 138-140, 144,
    186, 197, 198, 204, 206
Generative Design Grammar, 102, 138,
    140
Gothic, 46
Graph grammars, 23, 24
Grasshopper, 19, 20, 32, 34, 104, 108,
    154, 155, 205, 209, 211, 214, 218, 220,
    232
Group Board, 127
GuidiGo, 43

H

Haptipedia, 15
Hillier, 103, 147, 148, 163-165

I

Indus Valley Civilisation, 10
Inequality genotype, 147, 148, 153-156
Integration, 13, 32, 33, 148, 150, 152,
    165, 167-171, 173, 186, 205, 207
Intelligibility, 164, 165, 167, 169, 170
Isovist area, 166, 167, 171-173
Ivan Sutherland, 12

J

Jaggedness, 165-167, 170-172
Java, 32, 144

K

KPF, 32
Kubity Pro, 42

L

L-systems, 23, 24
Liuyuan, 147, 161-163
Liuyuan garden, 152, 153, 160
Luigi Moretti, 31

M

MaxScript, 19
Maya, 19, 42
Mean depth, 148, 152, 153, 165
Mimio, 128
Moretti, Luigi, 31

N

Neutra, Richard, 131
NURBs, 13, 31, 32

O

ObjectArx, 19
Occlusivity, 166, 167, 170-172
OpenGL, 18

P

Palladian grammar, 138
Parametricism, 33, 81 PIVOT, 44
Prairie Houses grammar, 138
Preview, 19, 35, 42, 212, 218-221
Processing, 12, 14, 17, 21, 131
Python, 32
Python script, 32

R

Renaissance, 31
Rhino, 19, 20, 32, 42, 104, 108, 209, 214,
    218, 219, 224
RoMA, 45
Ruby, 32

S

Sagrada Familia, 31
Sensorama, 46
Serlio, 31
Shape grammars, 22-24, 26, 138, 140,
    141

SightSpace, 42
SightSpace Pro 3D, 42
SiGraDi, 33
Siza Houses grammar, 138
Sketchpad, 12, 13, 16, 17
Sketchup, 42, 127, 132, 138
Smart Board, 128
SoftImage, 19
Soho Shang Du, 33
Soho Shang Du building, 33
Space syntax, 7, 103, 147, 148, 154, 164, 165, 174, 187
SPAN, 32
Step depth, 148
StreetMuseum, 44
Studierstube, 41
Swarm intelligence, 25, 26

T

Tangible user interface, 15, 70
The Sword of Damocles, 46
Tobbi Studio, 73
Total depth, 148, 149, 165
TX series, 12

U

UCL Depthmap, 149, 168
Umbra3D, 42, 43

Unity 3D, 43
UNStudio, 32

V

VAS, 73
Venice Biennial, 33
VGA map, 168, 170
VirtualBrick, 15
Virtual Reality (VR), 4, 15, 35, 45, 48, 127, 205
Visual attention simulation, 73
Visual Basic, 19, 32
Vitruvius, 31
VTT, 44
Vuforia, 43

Y

Yuyuan, 147, 162
Yuyuan garden, 103, 146, 149, 150-153, 157, 158, 161, 163, 166-174, 187

Z

Zaha Hadid Architects, 32
Zhuozhengyuan, 147
Zhuozhengyuan garden, 152, 153, 159